AF510784

PHYSICS RESEARCH AND TECHNOLOGY

RECENT ADVANCES IN QUARKS RESEARCH

PHYSICS RESEARCH AND TECHNOLOGY

Additional books in this series can be found on Nova's website under the Series tab.

Additional e-books in this series can be found on Nova's website under the e-book tab.

PHYSICS RESEARCH AND TECHNOLOGY

RECENT ADVANCES IN QUARKS RESEARCH

HARUKI FUJIKAGE
AND
KYOU HYOBANSHI
EDITORS

nova publishers
New York

Copyright © 2013 by Nova Science Publishers, Inc.

All rights reserved. No part of this book may be reproduced, stored in a retrieval system or transmitted in any form or by any means: electronic, electrostatic, magnetic, tape, mechanical photocopying, recording or otherwise without the written permission of the Publisher.

For permission to use material from this book please contact us:
Telephone 631-231-7269; Fax 631-231-8175
Web Site: http://www.novapublishers.com

NOTICE TO THE READER

The Publisher has taken reasonable care in the preparation of this book, but makes no expressed or implied warranty of any kind and assumes no responsibility for any errors or omissions. No liability is assumed for incidental or consequential damages in connection with or arising out of information contained in this book. The Publisher shall not be liable for any special, consequential, or exemplary damages resulting, in whole or in part, from the readers' use of, or reliance upon, this material. Any parts of this book based on government reports are so indicated and copyright is claimed for those parts to the extent applicable to compilations of such works.

Independent verification should be sought for any data, advice or recommendations contained in this book. In addition, no responsibility is assumed by the publisher for any injury and/or damage to persons or property arising from any methods, products, instructions, ideas or otherwise contained in this publication.

This publication is designed to provide accurate and authoritative information with regard to the subject matter covered herein. It is sold with the clear understanding that the Publisher is not engaged in rendering legal or any other professional services. If legal or any other expert assistance is required, the services of a competent person should be sought. FROM A DECLARATION OF PARTICIPANTS JOINTLY ADOPTED BY A COMMITTEE OF THE AMERICAN BAR ASSOCIATION AND A COMMITTEE OF PUBLISHERS.

Additional color graphics may be available in the e-book version of this book.

LIBRARY OF CONGRESS CATALOGING-IN-PUBLICATION DATA

Recent advances in quarks research / [edited by] Haruki Fujikage and Kyou Hyobanshi.
pages cm
Includes bibliographical references and index.
ISBN 978-1-62257-970-9 (hardcover)
1. Quarks. I. Fujikage, Haruki. II. Hyobanshi, Kyou.
QC793.5.Q252R43 2013
539.7'2167--dc23

2012035887

Published by Nova Science Publishers, Inc. † New York

CONTENTS

PREFACE

In this book, the authors present current advances in quarks research. Topics discussed in this compilation include the self-interaction mass formula relating all leptons and quarks to the electron; dirac model extension for finite size leptons and quarks in (10+1)D spacetime; quark confinement mechanisms for baryons; thermodynamical description of hadron-quark phase transition and its implications on compact-star phenomena; and the quark confinement analogy with baryonic bound states in superconductors.

In Chapter 1, an accurate empirical self-interaction mass formula has been obtained for all twelve leptons and quarks, in which the baryon number B and the charge number Q enter as eigenvalue surrogates for the strong and the electromagnetic self-interactions. All lepton and quark masses appear directly related to the electron mass.

In a Dirac model extension to 10 spatial dimensions (3 external plus 7 internal), leptons and quarks at rest are viewed to have a universal size and prolate spheroid shape, with mass and charge differences wholly attributed to their energy and charge densities through the universal spheroid. It is noted in Chapter 2 that simple volumetric quantum conditions on the 10D prolate spheroid and on its 4D prolate spheroid projection imply the fine-structure constant theoretical value $\alpha = (137.0360824)^{-1}$.

The confinement mechanism proposed earlier and then successfully applied to meson spectroscopy by the author is extended over baryons. For this aim the wave functions of baryons are built as tensorial products of those corresponding to the 2-body problem underlying the confinement mechanism of two quarks. This allows one to obtain the Hamiltonian of the quark interactions in a baryon and, accordingly, the possible energy spectrum of the latter. Also one may construct the electric and magnetic form factors of baryon in a natural way which entails the expressions for the root-mean-square radius and anomalous magnetic moment. To illustrate the formalism in Chapter 3 for the sake of simplicity only symmetrical baryons (i.e., composed from three quarks of the same flavours) $\Delta^{++}, \Delta, \Omega$ are considered. For them the masses, the rootmean- square radii and anomalous magnetic moments are expressed in an explicit analytical form through the parameters of the confining SU(3)-gluonic field among quarks and that enables one to get a number of numerical estimates for the mentioned parameters from experimental data.

The authors also discuss chiral limit for the baryons under consideration and estimate the purely gluonic contribution to their masses. Further the problem of masses in particle physics is shortly discussed within the framework of the given approach. Finally, a few remarks are made about the so-called Yang-Mills Millennium problem and a possible way for proving it is outlined.

One of the most promising possibilities may be the appearance of quark matter in astrophysical phenomena in the light of recent progress in observations. The mechanism of deconfinement is not well understood, but the thermodynamical aspects of the hadron-quark (HQ) phase transition have been extensively studied in recent years. Then the mixed phase of hadron and quark matter becomes important; the proper treatment is needed to describe the HQ phase transition and derive the equation of state (EOS) for the HQ matter, based on the Gibbs conditions for phase equilibrium. The authors here adopt a EOS based on the baryon-baryon interactions including hyperons for the hadron phase, while they use rather simple EOS within the MIT bag model for the quark phase. One of the interesting consequences may be the appearance of the inhomogeneous structures called "pasta", which are brought about by the surface and the Coulomb interaction effects. The authors present in Chapter 4 a comprehensive review of our recent works about the HQ phase transition in various astrophysical situations: cold catalyzed matter, hot matter and neutrino-trapped matter. The authors show how the pasta structure becomes unstable by the charge screening of the Coulomb interaction, thermal effect or the neutrino trapping effect. Such inhomogeneous structure may affect astrophysical phenomena through its elasticity or thermal properties. Here the authors also discuss some implications on supernova explosion, gravitational wave and cooling of compact stars.

As discussed in Chapter 5, there are profound analogies between ideas in particle physics and the corresponding ones in condensed-matter physics. There is also an interesting analogy between quarks and fractionally quantized-flux vortices in superconductors. Half-quantized flux vortices exist in two-component superconductors under a magnetic field. They can be interpreted as a monopole, and two half-quantum flux vortices form a bound state connected by a domain wall. The authors can formulate vortices with 1/3-quantum flux in three-component superconductors with equivalent three bands. Under the duality transformation between charge and magnetic flux, vortices with 1/3-quantum magnetic flux can be regarded as quarks with the 1/3 charge. A three-vortex bound state is formulated and this bound state resembles a baryon in QCD. The authors have also a two-vortex bound state, formed by vortices with 1/3 and -1/3 quantized flux, respectively, that corresponds to a meson. Two or three fractionally quantized vortices are connected by domain walls and form a bound state. The quark confinement is discussed by analogy with baryonic bound states in superconductors.

In: Recent Advances in Quarks Research
Editors: H. Fujikage and K. Hyobanshi

ISBN: 978-1-62257-970-9
© 2013 Nova Science Publishers, Inc.

Chapter 1

SELF-INTERACTION MASS FORMULA THAT RELATES ALL LEPTONS AND QUARKS TO THE ELECTRON

Gerald Rosen[*]
Department of Physics, Drexel University, Philadelphia, US

Abstract

An accurate empirical self-interaction mass formula has been obtained for all twelve leptons and quarks, in which the baryon number B and the charge number Q enter as eigenvalue surrogates for the strong and the electromagnetic self-interactions. All lepton and quark masses appear directly related to the electron mass.

PACS. 12.15. Ff – Quark and lepton models – specific theories – particle systematics

Recently very detailed high-precision calculations have established the masses of the up and down quarks and as

[*] E-mail address: gerald.h.rosen@drexel.edu

$$m_u = 2.01^{(10)\ \text{MeV}} \qquad m_d = 4.77^{(15)\ \text{MeV}} \tag{1}$$

at the most appropriate 2 GeV energy-scale in the $\overline{\text{MS}}$ renormalization scheme [1]. Thus the masses of the first-generation leptons and quarks are given by the *self-interaction mass formula*

$$m = m_e \langle (8\mathbf{B}-\mathbf{Q})^2 \rangle \tag{2}$$

in which m_e is the experimental mass of the electron (taken as empirical input) and **B**, **Q** are commuting quantum operators that give the baryon number $B \equiv \langle \mathbf{B} \rangle$ (eigenvalue surrogate for the strong interaction) and the charge number $Q \equiv \langle \mathbf{Q} \rangle$ (eigenvalue surrogate for the electromagnetic interaction), with $\langle \mathbf{BQ} \rangle = BQ$, $\langle \mathbf{B}^2 \rangle = B^2$, and $\langle \mathbf{Q}^2 \rangle = Q^2 + \varepsilon$ where $\varepsilon = 2.97 \times 10^{-11}$ derives from the virtual charge-distribution around a neutrino due to electroweak closed-loop processes [2, 3]. Hence from (2) we obtain

$$\begin{aligned} m_{\nu_1} &= \varepsilon\, m_e = 1.52 \times 10^{-5}\ \text{eV} \\ m_e &= 0.510999\ \text{MeV} \\ m_u &= 2.044\ \text{MeV} \\ m_d &= 4.599\ \text{MeV} \end{aligned} \tag{3}$$

Comparing (1) and (3), we see that the empirical formula (2) yields accurate values for the u and d masses. Lepton and quark self-interaction mass given by (2) for the first generation has an absolute physical character, energy-scale independent like pole mass [4] but with a physical meaning that does not involve perturbation-renormalization theory[†] (†).

† Existence of a Higgs boson to break the electroweak symmetry and engender the zero-mass photon and the massive $W^{\pm}$ and Z° bosons is of course likely. However, it is uncertain whether the

The self-interaction mass formula (2) can be generalized for the second and third generations. Let $\xi = \xi^3 = (0, -1, +1)$ be the *generation quantum number*, with $\xi = 0$ for the first, $\xi = -1$ for the second, and $\xi = +1$ for the third. By employing data-fitting techniques [6,7], we find

$$m = m_e \langle (n\mathbf{B}-\mathbf{Q})^2 \rangle \, 3^{|\xi|} (4.1)^{4|\xi|+\xi} \tag{4}$$

in which the even-integer valued coefficient of **B** is

$$n = 2^{3+(P+\frac{1}{2})\xi+(2P-\frac{3}{2})|\xi|} \, 3^{P(|\xi|-\xi)/2} \tag{5}$$

Appearing in (5) is the *projection quantum number* $P = P^2 \equiv |B+Q|$ which equals zero for $|Q| = 0$ or $\frac{1}{3}$ and equals unity for $|Q| = \frac{2}{3}$ or 1. Hence (5) reduces to

$$n = 2^{3+(\xi-3|\xi|)/2} \quad \text{for } P=0$$

$$n = 2^{3+(3\xi+|\xi|)/2} \, 3^{(|\xi|-\xi)/2} \quad \text{for } P=1 \tag{6}$$

In the final term in (4), the base factor (4.1) = (41/10) appears for the mass scaling of the three generations, as in earlier empirical investigations that were restricted to the charged-leptons [8,9]. The logical simplicity and quantum-theoretic character of the mass formula (4) are clearly evident, with all lepton and quark masses manifestly related to that of the electron.

The quantum numbers for leptons and quarks appear with their self-interaction masses (4) in the Table.

Higgs should also generate lepton and quark masses through a hierarchy of Yukawa couplings [5]. While the four fundamental bosons H, γ, $W^{\pm}$ and Z° most probably relate to each other, the twelve fundamental fermions appear to relate through the self-interactions mass formula (4). The lepton and quark mass values subsequently enter the standard model Lagrangian as pre-established input, with the mixing of quark states, as well as the mixing of neutrino states, viewed to be a standard model field-interaction phenomenon.

The overall agreement with the experimental masses [10,11] is quite striking. In particular, we have the fractional deviations $\delta m_\mu / m_\mu = -2.56 \times 10^{-5}$ and $\delta m_\tau / m_\tau = -7.32 \times 10^{-5}$.

Table. Quantum members and self-interaction masses for the leptons and quarks

	ξ	B	Q	P	n	m
ν_1	0	0	0	0	8	1.52×10^{-5} eV
e	0	0	–1	1	8	0.510999
u	0	⅓	⅔	1	8	2.044
d	0	⅓	–⅓	0	8	4.599
ν_2	–1	0	0	0	2	3.14×10^{-3} eV
μ	–1	0	–1	1	12	105.6557
c	–1	⅓	⅔	1	12	1173.95
s	–1	⅓	–⅓	0	2	105.6557
ν_3	1	0	0	0	4	5.28×10^{-2} eV
τ	1	0	–1	1	32	1776.07
t	1	⅓	⅔	1	32	177,607
b	1	⅓	–⅓	0	4	4933.53

Masses from (4) are in MeV except for the neutrinos, where eV magnitudes are indicated.

Acknowledgments

The author would like to thank Christine Davies for several illuminating communications regarding the work of her group [1]. At Drexel, Margaret Dominy helped with a literature survey, and Jacqueline Sampson assisted with the manuscript preparation.

References

[1] McNeile C. et al., arXiv: 1004.4285, v1 (2010).
[2] Bernabéu J. et al., *Phys. Rev. Lett.*, 89 (2002) 101802.
[3] Fujikawa K. and SHROCK R., *Phys. Rev. D*, 69 (2004) 013007.
[4] Kronfeld A.S., *Phys. Rev. D*, 58 (1998) 051501.

[5] Babu K.S. et al., *Phys. Rev. Lett.,* 67 (1993) 545.
[6] Tarantola A., *Inverse Problem Theory - Methods for Data Fixing and Model Parameter Estimation* (Elsevier, Amsterdam) 1987.
[7] Vapnik V.N., *Statistical Learning Theory* (Wiley, New York) 1998.
[8] Sirlin A., *Comm. Nucl. Part. Phys.,* 21 (1994) 227.
[9] Rosen G., *Europhys. Lett.,* 62 (2003) 473.
[10] Nakamura K. et al., *J. Phys. G,* 37 (2010) 075021.
[11] Particle Data Group (2012) http://pdq.lbl.gov/.

In: Recent Advances in Quarks Research
Editors: H. Fujikage and K. Hyobanshi
ISBN: 978-1-62257-970-9
© 2013 Nova Science Publishers, Inc.

Chapter 2

DIRAC MODEL EXTENSION FOR FINITE-SIZE LEPTONS AND QUARKS IN (10+1)D SPACETIME: QUANTUM CONDITIONS THAT IMPLY α^{-1}=137.0360824

Gerald Rosen*
Department of Physics, Drexel University, Philadelphia, PA, US

Abstract

In a Dirac model extension to 10 spatial dimensions (3 external plus 7 internal), leptons and quarks at rest are viewed to have a universal size and prolate spheroid shape, with mass and charge differences wholly attributed to their energy and charge densities through the universal spheroid. It is noted that simple volumetric quantum conditions on the 10D prolate spheroid and on its 4D prolate spheroid projection imply the fine-structure constant theoretical value $\alpha = (137.0360824)^{-1}$.

PACS. 12.90. +b – Miscellaneous theoretical ideas and models

* E-mail address: gerald.h.rosen@drexel.edu or gr@geraldrosen.com

Half a century ago, Dirac [1] proposed a finite-size spherical model for an electron at rest, a model which derived in part from the conjectured existence of a fundamental length ℓ. The Dirac model has been revisited with elaborations more recently [2–5], subject to the contemporary awareness that ℓ must be smaller thanabout 1×10^{-18} cm, a size compatible with the point-particle approximation of QED and the standard model for leptons and quarkswith an indication of finite-size still below current experimental resolution. Like crystals forming in a fluid medium, leptons and quarks are viewed as structured entities within their respective (motion and interaction) controlling quantum fields.

Here we consider a Dirac model extension in string-theory inspired (10+1)Dspacetime. Without conjecturing the magnitude of ℓ, we choose the fundamental length unit to be ℓ, and employ physical units that make $\ell=1$, along with $\hbar=1$and $c=1$. We investigate the most natural universal shape of leptons and quarks in 10 spatial dimensions for particles that are spherical at rest in the 3D external space cross-section. By requiring symmetry between the external (x_1,x_2,x_3) space and the internal $(x_4,x_5,x_6)$and (x_7,x_8,x_9) spaces, we are led to consider the 10D prolate spheroid

$$R_{10}\equiv\left\{\ x_1,\dots,x_9,x_{10}:\ \left(a^{-2}\sum_{i=1}^{9}x_i^2+b^{-2}x_{10}^2\right)\leqq 1\ \right\} \qquad (1)$$

with the volume

$$V_{10}=\int_{R_{10}}\left(\prod_{i=1}^{9}dx_i\right)dx_{10}=\frac{\pi^5}{5!}a^9b \qquad (2)$$

In (1) and (2), *a*is the semi-minor axis, as well as the radius of the lepton or quark in the external (x_1,x_2,x_3)space, while *b* is the semi-major axis in the internal x_{10} direction, the axis of revolution of the prolate spheroid, possibly associated with lepton-quark spin. Leptons and quarks at rest are viewed to have the universal shape given by (1), with their physical size deferred to future experimental determination of ℓ, and with their mass and charge differences attributed to different energy and charge densities through their R_{10} volume.

Next, let us associate local electromagnetic gauge transformations with x_{10} and *b*, as follows. Wavefunctions of charged particles are modified by a phase factor under local electromagnetic gauge transformations:

$$\psi \xrightarrow{\gamma} (\exp(i\varepsilon\gamma))\ \psi \tag{3}$$

where ε denotes the particle's charge and γ is the gauge parameter. Since the basic unit of electric charge is $\varepsilon = e/3$ and all lepton and quark charges are positive or negative multiples thereof, all gauge transformations are contained in the range $-b \leqq \gamma \leqq b$ with

$$b \equiv 3\pi / e \tag{4}$$

We identify the internal coordinate x_{10} to in fact be γ, $x_{10} \equiv \gamma$, and concomitantly assign *b* to have the physical value (4) in (1) and (2).

Now let $P = P^2$ denote the projection operator that for $P = 0$ maps the 10D prolate spheroid (1) into the 4D prolate spheroid

$$R_4 \equiv \left\{ x_1, x_2, x_3, x_{10} : \left(a^{-2} \sum_{i=1}^{3} x_i^2 + b^{-2} x_{10}^2 \right) \leqq 1 \right\} \tag{5}$$

with the volume

$$V_4 = \int_{R_4} dx_1 dx_2 dx_3 dx_{10} = \frac{\pi^2}{2} a^3 b \tag{6}$$

Note that (1) and (5) are subsumed by the single definition

$$R_{4+6P} \equiv \{ x_1, x_2, x_3, Px_4, \ldots Px_9, x_{10} : \left(a^{-2} \sum_{i=1}^{3+6P} x_i^2 + b^{-2} x_{10}^2 \right) \leqq 1 \} \tag{7}$$

Viewing P as a quantum operator with eigenvalues $P = 0$ and 1, we seek the simplest mathematically compatible volumetric quantum conditions for (2) and (6). We soon find

$$V_{4+6P} = \frac{1}{4} a^{2-3P} b^{2+P}, \tag{8}$$

which in combination with (2) and (6) yields

$$\frac{\pi^5}{5!} a^9 b = \frac{1}{4} a^{-1} b^3 \quad \text{and} \quad \frac{\pi^2}{2} a^3 b = \frac{1}{4} a^2 b^2. \tag{9}$$

Solving eqs. (9), we obtain

$$a = (120/\pi)^{1/8} = 1.576717623 \tag{10}$$

$$b = 2\pi^2 a = 31.12315838 \tag{11}$$

From (11) and (4) it follows that

$$e = 3\pi / b = 0.3028220288 \tag{12}$$

Note that the left-side members of (11) and (12) imply

$$(2\pi a)(e/3) = 1, \tag{13}$$

which states that the lepton or quark circumference in the external (x_1, x_2, x_3) space, $2\pi a$, is the reciprocal of the basic charge unit (e/3). Finally, (12) yields the fine-structure constant value

$$\alpha = e^2 / 4\pi = (137.0360824)^{-1}. \tag{14}$$

The latter theoretical value is less than the CODATA experimental value, $\alpha_{exp} = (137.0359991)^{-1}$, by a mere 4.5×10^{-9}.

References

[1] Dirac P.A.M., *Proc. Roy. Soc. (London) A*, 268 (1962) 37.
[2] Goldhaber M., *Proc. Natl. Acad. Sci.*, 99 (2002) 33.
[3] Rosen G., *EPL*, 62 (2003) 473.
[4] Goldhaber M., *Int. J. Mod. Phys.A (Suppl.),* 19 (2004) 167.
[5] Rosen G., *Mod. Phys. Lett. A*,22 (2007) 283.

In: Recent Advances in Quarks Research ISBN: 978-1-62257-970-9
Editors: H. Fujikage and K. Hyobanshi © 2013 Nova Science Publishers, Inc.

Chapter 3

QUARK CONFINEMENT MECHANISM FOR BARYONS

Yu. P. Goncharov*
Theoretical Group, Experimental Physics Department,
State Polytechnical University, Sankt-Petersburg, Russia

Abstract

The confinement mechanism proposed earlier and then successfully applied to meson spectroscopy by the author is extended over baryons. For this aim the wave functions of baryons are built as tensorial products of those corresponding to the 2-body problem underlying the confinement mechanism of two quarks. This allows one to obtain the Hamiltonian of the quark interactions in a baryon and, accordingly, the possible energy spectrum of the latter. Also one may construct the electric and magnetic form factors of baryon in a natural way which entails the expressions for the root-mean-square radius and anomalous magnetic moment. To ullustrate the formalism in the given Chapter for the sake of simplicity only symmetrical baryons (i.e., composed from three quarks of the same flavours) Δ^{++}, Δ^{-}, Ω^{-} are considered. For them the masses, the root-mean-square radii and anomalous magnetic moments are expressed in an explicit analytical form through the parameters of the confining SU(3)-gluonic field among quarks and that enables one to get a number of numerical estimates for the mentioned parameters from experimental data.

*E-mail address: ygonch@chern.hop.stu.neva.ru

We also discuss chiral limit for the baryons under consideration and estimate the purely gluonic contribution to their masses. Further the problem of masses in particle physics is shortly discussed within the framework of the given approach. Finally, a few remarks are made about the so-called Yang-Mills Millennium problem and a possible way for proving it is outlined.

PACS. 12.38.-t, 12.38.Aw, 14.20.-c

Keywords: Quantum chromodynamics, Quark confinement, Baryons

1. Introduction

The present Chapter is a natural continuation of our previous papers on studying both heavy quarkonia and the pseudoscalar meson nonet within the framework of the (quark) confinement mechanism proposed earlier by the author. Global strategy may consist in reconsidering the whole spectroscopy of both mesons and baryons from the positions of the mentioned mechanism so, e.g., exploring the pseudoscalar meson nonet is in essence only the first step in the given direction. But an extension of the approach over baryons requires a relativistic 3-body problem to be considered in contrast to mesons where we deal with a relativistic 2-body problem as follows from the standard quark model (SQM). It is clear, however, that the confinement mechanism under discussion (which in essence describes the mentioned relativistic 2-body problem) should also occupy a fitting place in the 3-body constructions. Let us, therefore, shortly outline the main features of the approach suggested.

In [1, 2, 3] for the Dirac-Yang-Mills system derived from QCD-Lagrangian an unique family of compatible nonperturbative solutions was found and explored, which could pretend to decsribing confinement of two quarks. The applications of the family to the description of both the heavy quarkonia spectra [4, 5, 6, 7] and a number of properties of pions, kaons, η- and η'-mesons [8, 9, 10, 11, 12, 13] showed that the confinement mechanism is qualitatively the same for both light mesons and heavy quarkonia. At this moment it can be decribed in the following way.

The next main physical reasons underlie linear confinement in the mechanism under discussion. The first one is that gluon exchange between quarks is realized with the propagator different from the photon-like one, and existence and form of such a propagator is a *direct* consequence of the unique confining nonperturbative solutions of the Yang-Mills equations [2, 3]. The second reason is that, owing to the structure of the mentioned propagator, quarks mainly emit and interchange the soft gluons so the gluon condensate (a classical gluon field) between quarks basically consists of soft gluons (for more details see Refs. [2, 3]) but, because of the fact that any gluon also emits gluons (still softer), the corresponding gluon concentrations rapidly become huge and form a linear confining magnetic colour field of enormous strengths, which leads to confinement of quarks. This is by virtue of the fact that just the magnetic part of the mentioned propagator is responsible for a larger portion of gluon concentrations at large distances since the magnetic part has stronger infrared singularities than the electric one. In the circumstances physically nonlinearity of the Yang-Mills equations effectively vanishes so the latter possess the unique nonperturbative confining solutions of the Abelian-like form (with the values in Cartan subalgebra of SU(3)-Lie algebra) [2, 3] which describe the gluon condensate under consideration. Moreover, since the overwhelming majority of gluons is soft they cannot leave the hadron (meson) until some gluons obtain additional energy (due to an external reason) to rush out. So we also deal with the confinement of gluons.

The approach under discussion equips us with the explicit wave functions for every two quarks (meson or quarkonium). The wave functions are parametrized by a set of real constants a_j, b_j, B_j describing the mentioned *nonperturbative* confining SU(3)-gluonic field (the gluon condensate) and they are *nonperturbative* modulo square integrable solutions of the Dirac equation in the above confining SU(3)-field and also depend on μ_0, the reduced mass of the current masses of quarks forming meson. It is clear that under the given approach just constants a_j, b_j, B_j, μ_0 determine all properties of any meson (quarkonium), i. e., the approach directly appeals to quark and gluonic degrees of freedom as should be according to the first principles of QCD. Also it is clear that the constants mentioned should be extracted from experimental data.

Such a program has been to a certain extent advanced in Refs. [4, 5, 6, 7, 8, 9, 10, 11, 12, 13] for both heavy quarkonia and the pseudoscalar meson nonet. It is clear, however, that baryons constitute no less physically important class of hadrons. But according to SQM (see, e. g., [17]) baryons are composed from three quarks, generally speaking, with different flavours. As a consequence, we are faced with a relativistic 3-body problem.

Under the circumstances the main aim of the present Chapter is to develop a certain approach to baryons based on the above confinement mechanism. For the sake of simplicity we shall here restrict ourselves to the symmetrical baryons (i.e., composed from three quarks of the same flavours) Δ^{++}, Δ^{-}, Ω^{-} inasmuch as in accordance with SQM $\Delta^{++} = uuu$, $\Delta^{-} = ddd$, $\Omega^{-} = sss$.

Section 2 contains a survey of main relations underlying description of any mesons (quarkonia) in our approach (the 2-body problem). Section 3 takes into account the considerations of Section 2 when constructing possible wave functions of baryons in our approach. Section 4 uses the obtained baryonic wave functions to build masses, the electric form factors, the root-mean-square radii $< R >$ and the anomalous magnetic moments of the symmetrical baryons under consideration in an explicit analytic form. Section 5 gives estimates for parameters of the confining SU(3)-gluonic field for baryons under discussion while Section 6 deals with discussion about chiral limit for baryons under studying. In Section 7 the problem of masses in particle physics is shortly discussed within the framework of approach to the chiral symmetry breaking in QCD proposed recently by the author. In Section 8 we make a few remarks about the so-called Yang-Mills Millennium problem and outline a possible way for proving it. Section 9 is devoted to a number of concluding remarks.

Appendices A and B contain the detailed description of main building blocks for meson wave functions in the approach under discussion, respectively: eigenspinors of the Euclidean Dirac operator on 2-sphere $\mathbb{S}^2$ and radial parts for the modulo square integrable solutions of Dirac equation in the confining SU(3)-Yang-Mills field while Appendix C recalls a few facts about tensorial products of Hilbert spaces necessary to construct the baryonic wave functions. At last, Appendix D supplements Section 2 with a proof of the uniqueness theorem from that Section in the case of SU(3)-Yang-Mills equations.

Further we shall deal with the metric of the flat Minkowski spacetime M that we write down (using the ordinary set of local spherical coordinates r, ϑ, φ for the spatial part) in the form

$$ds^2 = g_{\mu\nu} dx^\mu \otimes dx^\nu \equiv dt^2 - dr^2 - r^2(d\vartheta^2 + \sin^2\vartheta d\varphi^2) \,. \qquad (1)$$

Besides, we have $|\delta| = |\det(g_{\mu\nu})| = (r^2 \sin\vartheta)^2$ and $0 \le r < \infty$, $0 \le \vartheta < \pi$, $0 \le \varphi < 2\pi$.

Throughout the Chapter we employ the Heaviside-Lorentz system of units with $\hbar = c = 1$, unless explicitly stated otherwise, so the gauge coupling constant g and the strong coupling constant α_s are connected by the relation $g^2/(4\pi) = \alpha_s$. Further, we shall denote by $L_2(F)$ the set of the modulo square integrable complex functions on any manifold F furnished with an integration measure, then $L_2^n(F)$ will be the n-fold direct product of $L_2(F)$ endowed with the obvious scalar product while † and $*$ stand, respectively, for Hermitian and complex conjugation. Our choice of Dirac γ-matrices conforms to the so-called standard representation and is the same as in Ref. [8]. At last $\otimes$ means tensorial product of matrices and I_n is the unit $n \times n$ matrix so that, e.g., we have

$$I_3 \otimes \gamma^\mu = \begin{pmatrix} \gamma^\mu & 0 & 0 \\ 0 & \gamma^\mu & 0 \\ 0 & 0 & \gamma^\mu \end{pmatrix}$$

for any Dirac γ-matrix γ^μ and so forth.

When calculating we apply the relations $1\ \mathrm{GeV}^{-1} \approx 0.1973269679$ fm , $1\ \mathrm{s}^{-1} \approx 0.658211915 \times 10^{-24}$ GeV , $1\ \mathrm{V/m} \approx 0.2309956375 \times 10^{-23}\ \mathrm{GeV}^2$, $1\ \mathrm{T} = 4\pi \times 10^{-7}\mathrm{H/m} \times 1\ \mathrm{A/m} \approx 0.6925075988 \times 10^{-15}\ \mathrm{GeV}^2$, $\mu_N = 3.1524512326 \times 10^{-17}\ \mathrm{GeV/T} \approx 0.4552226197 \times 10^{-4}\ \mathrm{GeV}^{-1}$.

Finally, for the necessary estimates we shall employ the T_{00}-component (volumetric energy density) of the energy-momentum tensor for a SU(3)-Yang-Mills field which should be written in the chosen system of units in the form

$$T_{\mu\nu} = -F^a_{\mu\alpha}\, F^a_{\nu\beta}\, g^{\alpha\beta} + \frac{1}{4} F^a_{\beta\gamma}\, F^a_{\alpha\delta} g^{\alpha\beta} g^{\gamma\delta} g_{\mu\nu}\ . \tag{2}$$

2. Survey of Main Relations for the 2-Body Problem

2.1. Preliminaries

As was mentioned above, our considerations shall be based on the unique family of compatible nonperturbative solutions for the Dirac-Yang-Mills system (derived from QCD-Lagrangian) studied at the whole length in Refs. [1, 2, 3]. Referring for more details to those references, let us briefly decribe and specify only the relations necessary to us in the present necessary to us in the present necessary to us in the present Chapter.

One part of the mentioned family is presented by the unique nonperturbative confining solution of the SU(3)-Yang-Mills equations for the gluonic field $A = A_\mu dx^\mu = A^a_\mu \lambda_a dx^\mu$ (λ_a are the known Gell-Mann matrices, $\mu = t, r, \vartheta, \varphi$, $a = 1, ..., 8$) and looks as follows

$$\mathcal{A}_{1t} \equiv A^3_t + \frac{1}{\sqrt{3}} A^8_t = -\frac{a_1}{r} + A_1\ ,$$

$$\mathcal{A}_{2t} \equiv -A_t^3 + \frac{1}{\sqrt{3}}A_t^8 = -\frac{a_2}{r} + A_2 ,$$

$$\mathcal{A}_{3t} \equiv -\frac{2}{\sqrt{3}}A_t^8 = \frac{a_1 + a_2}{r} - (A_1 + A_2) ,$$

$$\mathcal{A}_{1\varphi} \equiv A_\varphi^3 + \frac{1}{\sqrt{3}}A_\varphi^8 = b_1 r + B_1 ,$$

$$\mathcal{A}_{2\varphi} \equiv -A_\varphi^3 + \frac{1}{\sqrt{3}}A_\varphi^8 = b_2 r + B_2 ,$$

$$\mathcal{A}_{3\varphi} \equiv -\frac{2}{\sqrt{3}}A_\varphi^8 = -(b_1 + b_2)r - (B_1 + B_2) \quad (3)$$

with the real constants a_j, A_j, b_j, B_j parametrizing the family. The word *unique* should be understood in the strict mathematical sense. In fact in Ref. [2] the following theorem was proved (see also appendix D):

The unique exact spherically symmetric (nonperturbative) solutions (depending only on r and r^{-1}) of SU(3)-Yang-Mills equations in Minkowski spacetime consist of the family of (3).

Additionally we impose the Lorentz condition on the sought solutions. The latter condition is necessary for quantizing the gauge fields consistently within the framework of perturbation theory (see, e. g. Ref. [41]), so we should impose the given condition that can be written in the form $\mathrm{div}(A) = 0$, where the divergence of the Lie algebra valued 1-form $A = A_\mu dx^\mu = A_\mu^a \lambda_a dx^\mu$ is defined by the relation (see, e. g., Refs. [40])

$$\mathrm{div}\, A = *(d * A) = \frac{1}{\sqrt{\delta}}\partial_\mu(\sqrt{\delta}g^{\mu\nu}A_\nu)$$

with the Hodge star operator $*$ (see Appendix D). It should be emphasized that, from the physical point of view, the Lorentz condition reflects the fact of transversality for gluons that arise as quanta of SU(3)-Yang-Mills field when quantizing the latter (see, e. g., Ref. [41]).

It should be noted that solution (3) was found early in Ref. [1] but its uniqueness was proved just in Ref. [2] (see also Ref. [3]). Besides, in Ref. [2] (see also Ref. [8]) it was shown that the above unique confining solutions (3) satisfy the so-called Wilson confinement criterion [14]. Up to now nobody contested the above results so if we want to describe interaction between quarks by spherically symmetric SU(3)-fields then they can be only those from the above theorem. On the other hand, the desirability of spherically symmetric (colour) interaction between quarks at all distances naturally follows from analysing the $p\bar{p}$-collisions (see, e.g., Ref. [16]) where one observes a Coulomb-like potential in events which can be identified with scattering quarks on each other, i.e., actually at small distances one observes the Coulomb-like part of solution (3). Under this situation, a natural assumption will be that the quark interaction remains spherically symmetric at large distances too but then, if trying to extend the Coulomb-like part to large distances in a spherically symmetric way, we shall inevitably come to the solution (3) in virtue of the above theorem.

Now one should say that the similar unique confining solutions exist for all semisimple and non-semisimple compact Lie groups, in particular, for SU(N) with $N \geq 2$ and U(N) with $N \geq 1$ [2, 3]. Explicit form of solutions, e.g., for SU(N) with $N = 2, 4$ can be found in Ref.[3] but it should be emphasized that components linear in r always represent the magnetic (colour) field in all the mentioned solutions. Especially the case U(1)-group is interesting which corresponds to usual electrodynamics. Under this situation, as was pointed out in Refs. [2, 3], there is an interesting possibility of indirect experimental verification of the confinement mechanism under discussion. Indeed the confining solutions of Maxwell equations for classical electrodynamics point out the confinement phase could be in electrodynamics as well. Though there exist no elementary charged particles generating a constant magnetic field linear in r, the distance from particle, after all, if it could generate this elecromagnetic field configuration in laboratory then one might study motion of the charged particles in that field. The confining properties of the mentioned field should be displayed at classical level too but the exact behaviour of particles in this field requires certain analysis of the corresponding classical equations of motion. Such a program has been recently realized in Ref. [46]. Motion of a charged (classical) particle was studied in the field representing magnetic part of the mentioned solution of Maxwell equations and it was shown that one deals with the full classical confinement of the charged particle in such a field: under any initial conditions the particle motion is accomplished within a finite region of space so that the particle trajectory is near magnetic field lines while the latter are compact manifolds (circles). Those results might be useful in thermonuclear plasma physics (for more details see [46]).

As has been repeatedly explained in Refs. [2, 3, 4, 5, 6, 8], parameters $A_{1,2}$ of solution (3) are inessential for physics in question and we can consider $A_1 = A_2 = 0$. Obviously we have $\sum_{j=1}^{3} \mathcal{A}_{jt} = \sum_{j=1}^{3} \mathcal{A}_{j\varphi} = 0$ which reflects the fact that for any matrix $\mathcal{T}$ from SU(3)-Lie algebra we have $\mathrm{Tr}\,\mathcal{T} = 0$. Also, as has been repeatedly discussed by us earlier (see, e. g., Refs. [2, 3]), from the above form it is clear that the solution (3) is a configuration describing the electric Coulomb-like colour field (components $A_t^{3,8}$) and the magnetic colour field linear in r (components $A_\varphi^{3,8}$) and we wrote down the solution (3) in the combinations that are just needed further to insert into the Dirac equation (4).

Another part of the family is given by the unique nonperturbative modulo square integrable solutions of the Dirac equation in the confining SU(3)-field of (3) $\Psi = (\Psi_1, \Psi_2, \Psi_3)$ with the four-dimensional Dirac spinors Ψ_j representing the jth colour component of the meson, so Ψ may describe the relative motion (relativistic bound states) of two quarks in mesons and the mentioned Dirac equation is written as follows

$$i\partial_t\Psi \equiv i\begin{pmatrix}\partial_t\Psi_1\\ \partial_t\Psi_2\\ \partial_t\Psi_3\end{pmatrix} = H\Psi \equiv$$

$$\begin{pmatrix}H_1 & 0 & 0\\ 0 & H_2 & 0\\ 0 & 0 & H_3\end{pmatrix}\begin{pmatrix}\Psi_1\\ \Psi_2\\ \Psi_3\end{pmatrix} = \begin{pmatrix}H_1\Psi_1\\ H_2\Psi_2\\ H_3\Psi_3\end{pmatrix}, \tag{4}$$

where the Hamiltonian H_j is

$$H_j = \gamma^0\left[\mu_0 - i\gamma^1\partial_r - i\gamma^2\frac{1}{r}\left(\partial_\vartheta + \frac{1}{2}\gamma^1\gamma^2\right) - \right.$$

$$\left. i\gamma^3\frac{1}{r\sin\vartheta}\left(\partial_\varphi + \frac{1}{2}\sin\vartheta\gamma^1\gamma^3 + \frac{1}{2}\cos\vartheta\gamma^2\gamma^3\right)\right] -$$

$$g\gamma^0\left(\gamma^0 A_{jt} + \gamma^3\frac{1}{r\sin\vartheta}A_{j\varphi}\right) \tag{5}$$

with the gauge coupling constant g while μ_0 is a mass parameter and one should consider it to be the reduced mass which is equal to $m_{q_1}m_{q_2}/(m_{q_1} + m_{q_2})$ with the current quark masses m_{q_k} (rest energies) of the corresponding quarks forming a meson (quarkonium).

Then the unique nonperturbative modulo square integrable solutions of (4) are at $j = 1, 2, 3$ (with Pauli matrix σ_1)

$$\Psi_j = e^{-i\omega_j t}\psi_j \equiv e^{-i\omega_j t} r^{-1}\begin{pmatrix}F_{j1}(r)\Phi_j(\vartheta,\varphi)\\ F_{j2}(r)\sigma_1\Phi_j(\vartheta,\varphi)\end{pmatrix} \tag{6}$$

with the 2D eigenspinor $\Phi_j = \begin{pmatrix} \Phi_{j1} \\ \Phi_{j2} \end{pmatrix}$ of the Euclidean Dirac operator $\mathcal{D}_0$ on the unit sphere $\mathbb{S}^2$, while the coordinate r stands for the distance between quarks. The explicit form of Φ_j is discussed in Appendix A. We can call the quantity ω_j the relative energy of the jth colour component of a meson (while ψ_j is the wave function of a stationary state for the jth colour component), but we can see that if we want to interpret (4) as an equation for eigenvalues of the relative motion energy, i. e., to rewrite it in the form $H\psi = \omega\psi$ with $\psi = (\psi_1, \psi_2, \psi_3)$ then we should put $\omega = \omega_j$ for any j so that $H_j\psi_j = \omega_j\psi_j = \omega\psi_j$. In this situation, if a meson is composed of quarks $q_{1,2}$ with different flavours then the energy spectrum of the meson will be given by $\epsilon = m_{q_1} + m_{q_2} + \omega$ with the current quark masses m_{q_k} (rest energies) of the corresponding quarks. On the other hand for determination of ω_j the following quadratic equation can be obtained [1, 2, 3]

$$[g^2a_j^2 + (n_j + \alpha_j)^2]\omega_j^2 - 2(\lambda_j - gB_j)g^2a_jb_j\,\omega_j +$$
$$[(\lambda_j - gB_j)^2 - (n_j + \alpha_j)^2]g^2b_j^2 - \mu_0^2(n_j + \alpha_j)^2 = 0\,, \tag{7}$$

which yields

$$\omega_j = \omega_j(n_j, l_j, \lambda_j) =$$
$$\frac{\Lambda_j g^2 a_j b_j \pm (n_j + \alpha_j)\sqrt{(n_j^2 + 2n_j\alpha_j + \Lambda_j^2)\mu_0^2 + g^2b_j^2(n_j^2 + 2n_j\alpha_j)}}{n_j^2 + 2n_j\alpha_j + \Lambda_j^2},$$
$$j = 1, 2, 3\,, \tag{8}$$

where $a_3 = -(a_1 + a_2)$, $b_3 = -(b_1 + b_2)$, $B_3 = -(B_1 + B_2)$, $\Lambda_j = \lambda_j - gB_j$, $\alpha_j = \sqrt{\Lambda_j^2 - g^2a_j^2}$, $n_j = 0, 1, 2, ...$, while $\lambda_j = \pm(l_j + 1)$ are the eigenvalues of Euclidean Dirac operator $\mathcal{D}_0$ on a unit sphere with $l_j = 0, 1, 2, ...$ It should be noted that in the papers [1, 2, 3, 4, 5, 6, 8] we used the ansatz (6) with the factor $e^{i\omega_j t}$ instead of $e^{-i\omega_j t}$ but then the Dirac equation (4) would look as $-i\partial_t\Psi = H\Psi$ and in equation (7) the second summand would have the plus sign while the first summand in numerator of (8) would have the minus sign. In the papers [9, 10] we returned to the conventional form of writing Dirac equation and this slightly modified the equations (7)–(8). In the given Chapter we conform to the same prescription as in Refs. [9, 10].

In line with the above we should have $\omega = \omega_1 = \omega_2 = \omega_3$ in energy spectrum $\epsilon = m_{q_1} + m_{q_2} + \omega$ for any meson (quarkonium) and this at once imposes two conditions on parameters a_j, b_j, B_j when choosing some experimental value for ϵ at the given current quark masses m_{q_1}, m_{q_2}.

The general form of the radial parts of (6) is considered in Appendix B. Within the given Chapter we need only the radial parts of (6) at $n_j = 0$ (the ground state) that are [see $(B.5)$]

$$F_{j1} = C_j P_j r^{\alpha_j} e^{-\beta_j r} \left(1 - \frac{gb_j}{\beta_j}\right), P_j = gb_j + \beta_j,$$

$$F_{j2} = iC_j Q_j r^{\alpha_j} e^{-\beta_j r} \left(1 + \frac{gb_j}{\beta_j}\right), Q_j = \mu_0 - \omega_j \tag{9}$$

with $\beta_j = \sqrt{\mu_0^2 - \omega_j^2 + g^2 b_j^2}$, while C_j is determined from the normalization condition $\int_0^\infty (|F_{j1}|^2 + |F_{j2}|^2) dr = \frac{1}{3}$. Consequently, we shall gain that $\Psi_j \in L_2^4(\mathbb{R}^3)$ at any $t \in \mathbb{R}$ and, as a result, the solutions of (6) may describe relativistic bound states (mesons) with the energy (mass) spectrum ϵ.

2.2. Nonrelativistic and the Weak Coupling Limits

It is useful to specify the nonrelativistic limit (when $c \to \infty$) for spectrum (8). For this one should replace $g \to g/\sqrt{\hbar c}$, $a_j \to a_j/\sqrt{\hbar c}$, $b_j \to b_j\sqrt{\hbar c}$, $B_j \to B_j/\sqrt{\hbar c}$ and, expanding (8) in $z = 1/c$, we shall get

$$\omega_j(n_j, l_j, \lambda_j) = \pm\mu_0 c^2 \left[1 \mp \frac{g^2 a_j^2}{2\hbar^2 (n_j + |\lambda_j|)^2} z^2\right]$$

$$+ \left[\frac{\lambda_j g^2 a_j b_j}{\hbar (n_j + |\lambda_j|)^2} \mp \mu_0 \frac{g^3 B_j a_j^2 f(n_j, \lambda_j)}{\hbar^3 (n_j + |\lambda_j|)^7}\right] z + O(z^2), \tag{10}$$

where $f(n_j, \lambda_j) = 4\lambda_j n_j (n_j^2 + \lambda_j^2) + \frac{|\lambda_j|}{\lambda_j} \left(n_j^4 + 6n_j^2 \lambda_j^2 + \lambda_j^4\right)$.

As is seen from (10), at $c \to \infty$ the contribution of linear magnetic colour field (parameters b_j, B_j) to the spectrum really vanishes and the spectrum in essence becomes the purely nonrelativistic Coulomb one (modulo the rest energy). Also it is clear that when $n_j \to \infty$, $\omega_j \to \pm\sqrt{\mu_0^2 + g^2 b_j^2}$. At last, one should specify the weak coupling limit of (8), i.e., the case $g \to 0$. As is not complicated to see from (8), $\omega_j \to \pm\mu_0$ when $g \to 0$. But then quantities $\beta_j = \sqrt{\mu_0^2 - \omega_j^2 + g^2 b_j^2} \to 0$ and wave functions of (9) cease to be the modulo square integrable ones at $g = 0$, i.e., they cease to describe relativistic bound states. Accordingly, this means that the equation (8) does not make physical meaning at $g = 0$.

We may seemingly use (8) with various combinations of signes ($\pm$) before the second summand in numerators of (8) but, due to (10), it is reasonable to take all signs equal to plus which is our choice within the Chapter. Besides, as is not complicated to see, radial parts in the nonrelativistic limit have the behaviour of form $F_{j1}, F_{j2} \sim r^{l_j+1}$, which allows one to call quantum number l_j angular momentum for the jth colour component though angular momentum is not conserved in the field (3) [1, 3]. So, for mesons under consideration we should put all $l_j = 0$.

2.3. Chiral Limit

There is one more interesting limit for relation (8) – the chiral one, i.e., the situation when $m_{q1}, m_{q2} \to 0$ which entails $\mu_0 \to 0$ and (8) reduces to (at $j = 1, 2, 3$)

$$(\omega_j)_{\rm chiral} = \frac{\Lambda_j g^2 a_j b_j \pm (n_j + \alpha_j) g |b_j| \sqrt{n_j^2 + 2n_j\alpha_j}}{n_j^2 + 2n_j\alpha_j + \Lambda_j^2}, \tag{8$'$}$$

which mathematically signifies that Dirac equation in the field (3) possesses a nontrivial spectrum of bound states even for massless fermions. Physically this gives us a possible approach to the problem of chiral symmetry breaking in QCD [12]: in chirally symmetric world masses of mesons are fully determined by the confining SU(3)-gluonic field between (massless) quarks and not equal to zero. Accordingly chiral symmetry is a sufficiently rough approximation holding true only when neglecting the mentioned SU(3)-gluonic field between quarks and no additional mechanism of the spontaneous chiral symmetry breaking connected to the so-called Goldstone bosons is required. As a result, e.g., masses of mesons from pseudoscalar nonet have a purely gluonic contribution and we have considered it in [12, 13] (see also sect. 6).

One can note that for to be the nonzero chiral limit of (8) the crucial role belongs to the colour magnetic field linear in r [parameters $b_{1,2}$ from solution (3)] inasmuch as chiral limit is equal exactly to zero when $b_{1,2} = 0$. On the contrary, when parameters $a_{1,2}$ the Coulomb colour electric part of solution (3) are equal to zero the chiral limit may be nonzero at $b_{1,2} \neq 0$, as is seen from

$(8')$ except for the case $n_j = 0$ when both parts of SU(3)-gluonic field (3) are important for confinement and mass generation in chiral limit.

2.4. Eigenspinors with $\lambda = \pm 1$

Finally it should be noted that spectrum (8) is degenerated owing to the degeneracy of eigenvalues for the Euclidean Dirac operator $\mathcal{D}_0$ on the unit sphere $\mathbb{S}^2$. Namely, each eigenvlalue of $\mathcal{D}_0$ $\lambda = \pm(l+1), l = 0, 1, 2...$, has multiplicity $2(l+1)$, so we has $2(l+1)$ eigenspinors orthogonal to each other. Ad referendum we need eigenspinors corresponding to $\lambda = \pm 1$ $(l = 0)$ so here is their explicit form [see $(A.16)$]

$$\lambda = -1 : \Phi = \frac{C}{2}\begin{pmatrix} e^{i\frac{\vartheta}{2}} \\ e^{-i\frac{\vartheta}{2}} \end{pmatrix} e^{i\varphi/2},$$

or

$$\Phi = \frac{C}{2}\begin{pmatrix} e^{i\frac{\vartheta}{2}} \\ -e^{-i\frac{\vartheta}{2}} \end{pmatrix} e^{-i\varphi/2},$$

$$\lambda = 1 : \Phi = \frac{C}{2}\begin{pmatrix} e^{-i\frac{\vartheta}{2}} \\ e^{i\frac{\vartheta}{2}} \end{pmatrix} e^{i\varphi/2},$$

or

$$\Phi = \frac{C}{2}\begin{pmatrix} -e^{-i\frac{\vartheta}{2}} \\ e^{i\frac{\vartheta}{2}} \end{pmatrix} e^{-i\varphi/2} \tag{11}$$

with the coefficient $C = 1/\sqrt{2\pi}$ (for more details, see Appendix A).

2.5. Meson Energy (Mass)

Within the present Chapter we shall use relations (8) at $n_j = 0 = l_j$ so energy (mass) of a meson is given by $\mu = m_{q_1} + m_{q_2} + \omega$ with $\omega = \omega_j(0, 0, \lambda_j)$ for any $j = 1, 2, 3$ whereas

$$\omega = \frac{g^2 a_1 b_1}{\Lambda_1} + \frac{\alpha_1 \mu_0}{|\Lambda_1|} = \frac{g^2 a_2 b_2}{\Lambda_2} + \frac{\alpha_2 \mu_0}{|\Lambda_2|}$$

$$= \frac{g^2 a_3 b_3}{\Lambda_3} + \frac{\alpha_3 \mu_0}{|\Lambda_3|} = \mu - m_{q_1} - m_{q_2} \tag{12}$$

and, as a consequence, the corresponding meson wave functions of (6) are represented by (9) and (11).

2.6. Choice of Quark Masses and the Gauge Coupling Constant

It is evident for employing the above relations we have to assign some values to quark masses and gauge coupling constant g. We take the current quark masses used in [9, 10, 12] and they are $m_u = 2.25$ MeV, $m_d = 5$ MeV, $m_s = 107.5$ MeV. Under the circumstances, the reduced mass μ_0 of (5) will be equal to $m_u/2$, $m_d/2$, $m_s/2$ accordingly. As to the gauge coupling constant $g = \sqrt{4\pi\alpha_s}$, it should be noted that recently some attempts have been made to generalize the standard formula for $\alpha_s = \alpha_s(Q^2) = 12\pi/[(33 - 2n_f)\ln(Q^2/\Lambda^2)]$ (n_f is number of quark flavours) holding true at the momentum transfer $\sqrt{Q^2} \to \infty$ to the whole interval $0 \le \sqrt{Q^2} \le \infty$. We shall employ one such a generalization used in Refs. [18]. It is written as follows ($x = \sqrt{Q^2}$ in GeV)

$$\alpha(x) = \frac{12\pi}{(33 - 2n_f)} \frac{f_1(x)}{\ln \frac{x^2 + f_2(x)}{\Lambda^2}} \tag{13}$$

with

$$f_1(x) =$$

$$1 + \left(\left(\frac{(1+x)(33-2n_f)}{12} \ln \frac{m^2}{\Lambda^2} - 1 \right)^{-1} + 0.6x^{1.3} \right)^{-1},$$

$$f_2(x) = m^2(1 + 2.8x^2)^{-2},$$

wherefrom one can conclude that $\alpha_s \to \pi = 3.1415...$ when $x \to 0$, i. e., $g \to 2\pi = 6.2831....$ We used (13) at $m = 1$ GeV, $\Lambda = 0.234$ GeV, $n_f = 3$, $x = 1232.0$ MeV or $x = 1672.45$ MeV to obtain $g \approx 3.367244706$ or $g \approx 2.841355055$ necessary for our further computations at the mass scale of Δ^{++}, Δ^-, or Ω^- respectively.

2.7. Electric Form Factor and the Root-Mean-Square Radius

For each meson (quarkonium) with the wave function $\Psi = (\Psi_j)$ of (6) we can define the electromagnetic current $J^\mu = \overline{\Psi}(I_3 \otimes \gamma^\mu)\Psi = (\Psi^\dagger\Psi, \Psi^\dagger(I_3 \otimes \alpha)\Psi) = (\rho, \mathbf{J})$, $\alpha = \gamma^0\gamma$. The electric form factor $f(K)$ is the Fourier transform of ρ

$$f(K) = \int \Psi^\dagger \Psi e^{-i\mathbf{K}\mathbf{r}} d^3x = \sum_{j=1}^{3} \int \Psi_j^\dagger \Psi_j e^{-i\mathbf{K}\mathbf{r}} d^3x =$$

$$\sum_{j=1}^{3} f_j(K) = \sum_{j=1}^{3} \int (|F_{j1}|^2 + |F_{j2}|^2)\Phi_j^\dagger \Phi_j \frac{e^{-i\mathbf{K}\mathbf{r}}}{r^2} d^3x,$$

$$d^3x = r^2 \sin\vartheta dr d\vartheta d\varphi \tag{14}$$

with the momentum transfer K. At $n_j = 0 = l_j$, as is easily seen, for any spinor of (11) we have $\Phi_j^\dagger \Phi_j = 1/(4\pi)$, so the integrand in (14) does not depend on φ and we can consider vector $\mathbf{K}$ to be directed along z-axis. Then $\mathbf{Kr} = Kr\cos\vartheta$ and with the help of (9) and relations (see Ref. [20]): $\int_0^\infty r^{\alpha-1} e^{-pr} dr = \Gamma(\alpha)p^{-\alpha}$, Re $\alpha, p > 0$, $\int_0^\infty r^{\alpha-1} e^{-pr} \begin{pmatrix} \sin(Kr) \\ \cos(Kr) \end{pmatrix} dr =$

$$\Gamma(\alpha)(K^2+p^2)^{-\alpha/2} \begin{pmatrix} \sin(\alpha\arctan(K/p)) \\ \cos(\alpha\arctan(K/p)) \end{pmatrix},$$

Re $\alpha > -1$, Re $p > |\mathrm{Im}\, K|$, $\Gamma(\alpha+1) = \alpha\Gamma(\alpha)$,

$$\int_0^\pi e^{-iKr\cos\vartheta} \sin\vartheta d\vartheta = 2\sin(Kr)/(Kr),$$

we shall obtain

$$f(K) = \sum_{j=1}^{3} f_j(K) =$$

$$\sum_{j=1}^{3} \frac{(2\beta_j)^{2\alpha_j+1}}{6\alpha_j} \cdot \frac{\sin[2\alpha_j \arctan(K/(2\beta_j))]}{K(K^2+4\beta_j^2)^{\alpha_j}}$$

$$= \sum_{j=1}^{3} \left(\frac{1}{3} - \frac{2\alpha_j^2 + 3\alpha_j + 1}{6\beta_j^2} \cdot \frac{K^2}{6} \right) + O(K^4), \tag{15}$$

wherefrom it is clear that $f(K)$ is a function of K^2, as should be, and we can determine the root-mean-square radius of meson (quarkonium) in the form

$$< r > = \sqrt{\sum_{j=1}^{3} \frac{2\alpha_j^2 + 3\alpha_j + 1}{6\beta_j^2}}. \tag{16}$$

When calculating (15) also the fact was used that by virtue of the normalization condition for wave functions we have $C_j^2[P_j^2(1-gb_j/\beta_j)^2 + Q_j^2(1+gb_j/\beta_j)^2] = (2\beta_j)^{2\alpha_j+1}/[3\Gamma(2\alpha_j+1)]$.

It is clear, we can directly calculate $< r >$ in accordance with the standard quantum mechanics rules as $< r >= \sqrt{\int r^2 \Psi^\dagger \Psi d^3x} = \sqrt{\sum_{j=1}^{3} \int r^2 \Psi_j^\dagger \Psi_j d^3x}$ and the result will be the same as in (16). So we should not call $< r >$ of (16) the *charge* radius of meson (quarkonium)– it is just the radius of meson (quarkonium) determined by the wave functions of (6) (at $n_j = 0 = l_j$) with respect to strong interaction, i.e., radius of confinement. Now we should note the expression (15) to depend on 3-vector $\mathbf{K}$. To rewrite it in the form holding true for any 4-vector Q, let us recall that according to general considerations (see, e.g., Ref. [21]) the relation (15) should correspond to the so-called Breit frame where $Q^2 = -K^2$ [when fixing the metric by (1)] so it is not complicated to rewrite (15) for arbitrary Q in the form

$$f(Q^2) = \sum_{j=1}^{3} f_j(Q^2) =$$

$$\sum_{j=1}^{3} \frac{(2\beta_j)^{2\alpha_j+1}}{6\alpha_j} \cdot \frac{\sin\left[2\alpha_j \arctan\left(\sqrt{|Q^2|}/(2\beta_j)\right)\right]}{\sqrt{|Q^2|}(4\beta_j^2 - Q^2)^{\alpha_j}} \tag{17}$$

which passes on to (15) in the Breit frame.

2.8. Magnetic Moment

We can define the volumetric magnetic moment density by $\mathbf{m} = q(\mathbf{r} \times \mathbf{J})/2 = q[(yJ_z - zJ_y)\mathbf{i} + (zJ_x - xJ_z)\mathbf{j} + (xJ_y - yJ_x)\mathbf{k}]/2$ with the meson charge q and $\mathbf{J} = \Psi^\dagger (I_3 \otimes \alpha)\Psi$. Using (6) we have in the explicit form

$$J_x = \sum_{j=1}^{3} (F_{j1}^* F_{j2} + F_{j2}^* F_{j1}) \frac{\Phi_j^\dagger \Phi_j}{r^2},$$

$$J_y = \sum_{j=1}^{3} (F_{j1}^* F_{j2} - F_{j2}^* F_{j1}) \frac{\Phi_j^\dagger \sigma_2 \sigma_1 \Phi_j}{r^2},$$

$$J_z = \sum_{j=1}^{3} (F_{j1}^* F_{j2} - F_{j2}^* F_{j1}) \frac{\Phi_j^\dagger \sigma_3 \sigma_1 \Phi_j}{r^2} \tag{18}$$

with Pauli matrices $\sigma_{1,2,3}$. Magnetic moment of meson (quarkonium) is $\mathbf{M} = \int_V \mathbf{m} d^3x$, where V is the volume of the meson (quarkonium) (the ball of radius $< r >$). Then at $n_j = l_j = 0$, as is seen from (9), (11), $F_{j1}^* = F_{j1}, F_{j2}^* = -F_{j2}$, $\Phi_j^\dagger \sigma_2 \sigma_1 \Phi_j = 0$ for any spinor of (11) which entails $J_x = J_y = 0$, i.e., $m_z = 0$ while $\int_V m_{x,y} d^3x = 0$ because of turning the integral over φ to zero, which is easy to check. As a result, the magnetic moments of mesons (quarkonia) with the wave functions of (6) (at $l_j = 0$) are equal to zero, as should be according to experimental data [17].

2.9. Magnetic Form Factor and Anomalous Magnetic Moment

Following Refs. [9, 10] we can, however, define magnetic form factor $F(K)$ of a meson if noticing that

$$\int_V \sqrt{m_x^2 + m_y^2 + m_z^2} d^3x \neq$$

$$\sqrt{\left(\int_V m_x d^3x\right)^2 + \left(\int_V m_y d^3x\right)^2 + \left(\int_V m_z d^3x\right)^2}$$

$$= M = 0\,.$$

Under the circumstances we should define $F(K)$ as the inverse Fourier transform of $\sqrt{m_x^2 + m_y^2 + m_z^2}$

$$\sqrt{m_x^2 + m_y^2 + m_z^2} = \frac{q}{(2\pi)^3} \int F(K) e^{i\mathbf{K}\mathbf{r}} d^3K\,,$$

so that

$$F(K) = \frac{1}{q} \int \sqrt{m_x^2 + m_y^2 + m_z^2} e^{-i\mathbf{K}\mathbf{r}} d^3x =$$

$$\frac{|q|}{q} \int r |J_z \sin\vartheta| e^{-i\mathbf{K}\mathbf{r}} d^3x =$$

$$\frac{1}{2\pi} \frac{|q|}{q} \int \sum_{j=1}^{3} |F_{j1}||F_{j2}| \frac{\sin^2\vartheta}{r} e^{-i\mathbf{K}\mathbf{r}} d^3x \tag{19}$$

for any spinor of (11) so the integrand in (19) does not again depend on φ. Then $\mathbf{K}\mathbf{r} = Kr\cos\vartheta$ and using the relation

$$\int_0^\pi e^{-iKr\cos\vartheta} \sin^3\vartheta d\vartheta =$$

$$4[\sin(Kr)/(Kr)^3 - \cos(Kr)/(Kr)^2],$$

we shall obtain

$$F(K) =$$

$$2\frac{|q|}{q}\int_0^\infty r\sum_{j=1}^{3} C_j^2 P_j Q_j r^{2\alpha_j} e^{-2\beta_j r}\left(1-\frac{g^2 b_j^2}{\beta_j^2}\right)\times$$

$$\left[\frac{\sin(Kr)}{(Kr)^3}-\frac{\cos(Kr)}{(Kr)^2}\right]dr$$

$$= \frac{1}{3}\frac{|q|}{q}\sum_{j=1}^{3}\frac{(2\beta_j)^{2\alpha_j+1}}{\alpha_j}\cdot\frac{\mathcal{P}_j\mathcal{Q}_j}{\mathcal{P}_j^2+\mathcal{Q}_j^2}\cdot\frac{1}{K^2(K^2+4\beta_j^2)^{\alpha_j}}\cdot$$

$$\left\{\frac{\sin[(2\alpha_j-1)\arctan(K/(2\beta_j))]}{(2\alpha_j-1)K(K^2+4\beta_j^2)^{-1/2}} - \cos[2\alpha_j \arctan(K/(2\beta_j))]\right\}$$

$$= \frac{2}{3}\frac{|q|}{q}\sum_{j=1}^{3}\frac{\beta_j}{\alpha_j}\cdot\frac{\mathcal{P}_j\mathcal{Q}_j}{\mathcal{P}_j^2+\mathcal{Q}_j^2}\left[\frac{\alpha_j(2\alpha_j+1)}{6\beta_j^2}-\right.$$

$$\left.\frac{\alpha_j(4\alpha_j^3+12\alpha_j^2+11\alpha_j+3)}{120\beta_j^4}K^2+O(K^4)\right]$$

$$= \frac{2}{3}\frac{|q|}{q}\sum_{j=1}^{3}\frac{2\alpha_j+1}{6\beta_j}\cdot\frac{\mathcal{P}_j\mathcal{Q}_j}{\mathcal{P}_j^2+\mathcal{Q}_j^2}\left(1-3\frac{2\alpha_j^2+5\alpha_j+3}{10\beta_j^2}\cdot\frac{K^2}{6}\right)+O(K^4) \quad (20)$$

with $\mathcal{P}_j = P_j(1-gb_j/\beta_j)$, $\mathcal{Q}_j = O_j(1+gb_j/\beta_j)$. It is clear from (20) that $F(K)$ is a function of K^2 and we can rewrite (20) for arbitrary 4-vector Q by the same manner as was done for electric form factor $f(K)$ in (17).

Now we can define *the anomalous magnetic moment* for a meson by the relation

$$\mu_a = F(K=0) = \frac{2}{3}\frac{|q|}{q}\sum_{j=1}^{3}\frac{2\alpha_j+1}{6\beta_j}\cdot\frac{\mathcal{P}_j\mathcal{Q}_j}{\mathcal{P}_j^2+\mathcal{Q}_j^2}, \quad (21)$$

as follows from (20). It should be noted that a possibility of existence of anomalous magnetic moments for mesons is permanently discussed in literature (see, e.g., Refs. [22, 23, 24] and references therein) though there is no experimental evidence in this direction [17]. Another matter is baryons with their nonzero anomalous magnetic moments and the latter are some essential characteristics of baryons. So that the relation (21) will be very useful to us below.

2.10. Remarks About the Relativistic Two-Body Problem

It is well known (see any textbook on quantum mechanics, e.g., Ref. [26]) that *nonrelativistic* two-body problem, when the particles interact with potential $V(\mathbf{r_1}, \mathbf{r_2})$ depending only on $|\mathbf{r_2} - \mathbf{r_1}| = r$, reduces to the motion of one particle with reduced mass $m = m_1 m_2/(m_1 + m_2)$ in potential $V(r)$, where r becomes the ordinary spherical coordinate from triplet (r, ϑ, φ). I.e., the bound states of *two* particles are the bound states of the particle with mass m in potential $V(r)$ and they are the modulo square integrable solutions of the corresponding Schrödinger equation. Another matter is *relativistic* two-body problem. As we emphasized in Refs. [2, 3], up to now it has no single-valued statement. But if noting that the most fundamental results of nonrelativistic quantum mechanics (hydrogen atom and so on) are connected with potentials $V(r)$ which are a part of the *electromagnetic* field $A = (A_0, \mathbf{A})$, i.e. $V(r) = A_0, \mathbf{A} = 0$ then one may propose some formulation of the relativistic two-body problem. Really, now $V(r)$ additionally obeys the Maxwell equations and if we want to generalize the corresponding Schrödinger equation to include an interaction with arbitrary electromagnetic field for $\mathbf{A} \neq 0$ then the answer is known: this is the Dirac equation with the replacement $\partial_\mu \to \partial_\mu - igA_\mu$, g is a gauge coupling constant. Indeed, when going back in nonrelativistic limit at the light velocity $c \to \infty$ the magnetic field $\mathbf{A}$ vanishes because, as is well known, in the world with $c = \infty$ there exist no magnetic fields (see any elementary textbook on physics). At the same time $V(r) = A_0$ does not vanish and remains the same as in nonrelativistic case and the Dirac equation turns into the Schrödinger equation (see Ref. [26]). But then we can see that m in the above Dirac equation should consider the same reduced mass as before since in nonrelativistic limit we again should come to the standard formulation of two-body problem through effective particle with reduced mass m. So we can draw the conclusion that if an electromagnetic field is a combination of electric $V(r) = A_0$ field between two charged elementary particles and some magnetic field $\mathbf{A} = \mathbf{A}(r)$ (which may be generated by the particles themselves and also depends only on r, distance between particles) then there are certain grounds to consider the given (quantum) relativistic two-body problem to be equivalent to the one of motion for one particle with usual reduced mass in the mentioned electromagnetic field. As a result, we can use the Dirac equation for finding possible relativistic bound states for such a particle implying that this is really some description of the corresponding two-body problem. Under this situation we should remark the following.

1. Although for simplicity we talked about electromagnetic field but everything holds true for any Yang-Mills field, in particular, for SU(3)-gluonic field while the Maxwell equations are replaced by the Yang-Mills ones.

2. There arises the question: whether the Maxwell or Yang-Mills equations possess solutions with spherically symmetric $A_0(r)$, $\mathbf{A}(r)$? The answer is given by the uniqueness theorem (see subsection 2.1 and Appendix D).

3. The Dirac equation in such a field has the nonperturbative spectrum (8) and the latter should be treated as the nonpeturbative *interaction energy* of two quarks and r as the distance between quarks and so on (see Section 2). So eq. (4) should be understood just in this manner. In line with the above we should have $\omega = \omega_1 = \omega_2 = \omega_3$ (with ω_j of (8)) in energy spectrum $\epsilon = m_{q_1} + m_{q_2} + \omega$ for any meson (quarkonium) and this at once imposes two conditions on parameters a_j, b_j, B_j of solution (3) when choosing some experimental value for ϵ at the given current quark masses m_{q_1}, m_{q_2}. At last, the Dirac equation in question is obtained from QCD lagrangian (with one flavor) if mass parameter in the latter is taken to be equal to the above reduced mass. Therefore, we can say that meson wave functions (6) are the nonperturbative solutions of the Dirac-Yang-Mills system directly derived from QCD-lagrangian.

4. To summarize, there exist good physical and mathematical grounds for formulation of the above relativistic two-body problem that has a correct non-relativistic limit and is Lorentz and gauge invariant. It is clear that all the above considerations can be justified only by comparison with experimental data but now we obtain some intelligible programme of further activity which has been partly realized in many our papers cited above.

Much of the above was disscussed in Refs. [2, 3] but perhaps in other words.

3. Wave Functions of Baryons

3.1. Preliminaries

In accordance with SQM [17] baryons are composed from three quarks, generally speaking, with different flavours, so a baryon can be represented in the schematic form shown in Fig. 1, where r_i are distances between the corresponding quarks. As a consequence, we are faced with a relativistic 3-body problem. Under the circumstances the main task is to obtain an interaction Hamiltonian H of the 3-quark system if taking that every two quarks interact through the

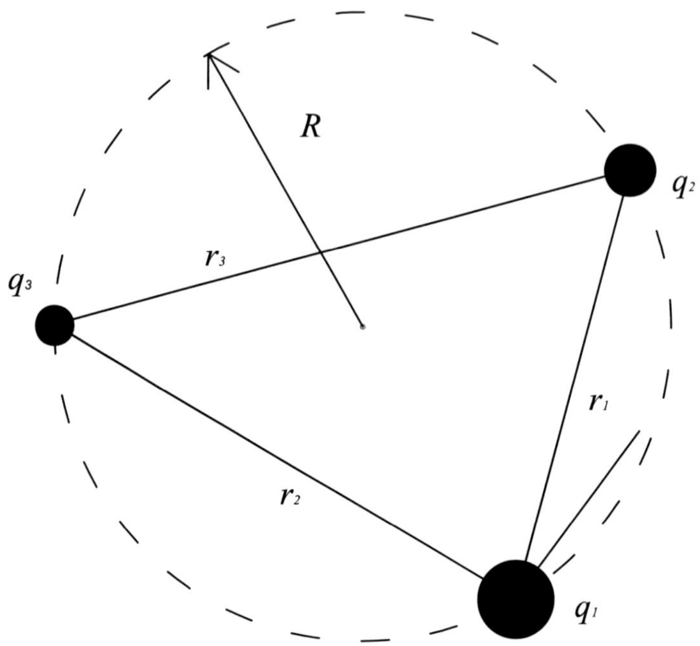

Figure 1. Schematic sketch of a baryon according to SQM.

gluonic field described by solution (3), in general, with different constants a_j, b_j, B_j for each pair of quarks. Then energy levels of a baryon will be given by relation $\epsilon = m_{q_1} + m_{q_2} + m_{q3} + \omega$ with the current quark masses m_{q_k} (rest energies) of the quarks constituting the baryon while the interaction energy ω should be subject to equation $H\psi = \omega\psi$ with a baryonic wave function ψ describing relative motion (bound state) of three quarks.

To construct wave functions of baryons let us at first make a remark concerning the picture of interaction between two quarks which was obtained when applying the confinement mechanism in question to description of heavy quarkonia and mesons from pseudoscalar nonet [4, 5, 7, 8, 9, 10, 11, 12, 13].

3.2. Qualitative Picture of Quark Interaction

Let us recall that, according to Refs. [3, 8], one can confront the field (3) with the T_{00}-component (the volumetric energy density of the SU(3)-gluonic field) of the energy-momentum tensor (2) so that

$$T_{00} \equiv T_{tt} = \frac{E^2 + H^2}{2} =$$

$$\frac{1}{2}\left(\frac{a_1^2 + a_1a_2 + a_2^2}{r^4} + \frac{b_1^2 + b_1b_2 + b_2^2}{r^2\sin^2\vartheta}\right) \equiv \frac{\mathcal{A}}{r^4} + \frac{\mathcal{B}}{r^2\sin^2\vartheta} \tag{22}$$

with electric E and magnetic H colour field strengths and with real $\mathcal{A} > 0, \mathcal{B} > 0$. One can also introduce magnetic colour induction $B = (4\pi \times 10^{-7}\mathrm{H/m})\,H$, where H in A/m.

We should now note that one can calculate energy $\mathcal{E}$ of gluon condensate conforming to solution (3) in a volume V through relation $\mathcal{E} = \int_V T_{00}r^2 \sin\vartheta dr d\vartheta d\varphi$ with T_{00} of (22) but one should take into account that classical T_{00} has a singularity along z-axis ($\vartheta = 0, \pi$) and we have to introduce some angle ϑ_0 so $\vartheta_0 \le \vartheta \le \pi - \vartheta_0$. As well as in Refs. [2, 8, 12], we may consider ϑ_0 to be a parameter determining some cone $\vartheta = \vartheta_0$ so the quark emits gluons outside of the cone. Now if there are two quarks Q_1, Q_2 and each of them emits gluons outside of its own cone $\vartheta = \vartheta_{1,2}$ (see Figs. 2, 3) then we have soft gluons (as mentioned in section 1) in regions I, II and between quarks.

Accordingly, we shall have some region V with gluon condensate between quarks Q_1, Q_2 and its vertical projection is shown in Fig. 2. Another projection of V onto a plane perpendicular to the one of Fig. 2 is sketched out in Fig. 3.

To clarify a physical meaning of the quantities $r_{1,2}$ in Figs. 2, 3, let us recall an analogy with classical electrodynamics where is well known (see, e. g., [25]) that the notion of classical electromagnetic field (a photon condensate) generated by a charged particle is applicable only at distances much greater than the Compton wavelength $\lambda_c = 1/m$ for the given particle with mass m. Within the QCD framework the parameter Λ_{QCD} plays a similar part (see, e.g., Ref. [17]). Namely, the notion of classical SU(3)-gluonic field (a gluon condensate) is not applicable at the distances much less than $1/\Lambda_{QCD}$. In accordance with subsection 2.6 we took $\Lambda_{QCD} = \Lambda = 0.234$ GeV which entails $1/\Lambda \sim 0.8433$ fm so one may consider $r_{1,2} \sim 0.08$ fm.

If taking into account that only the values of $\vartheta_{1,2}$, $\varphi_{1,2}$ between 0 and 90° are of physical meaning then it should be noted that the estimates of $\vartheta_{1,2}$, $\varphi_{1,2}$ obtained in Refs. [8, 12] show that $\vartheta_{1,2} \to \pi/2$, $\varphi_{1,2} \to 0$ with increasing the quark masses so gluon condensate (a classical SU(3)-gluonic field) is mainly concentrated near the axis connecting quarks. Under the circumstances we may represent a baryon composed from three pairs of the interacting quarks in the

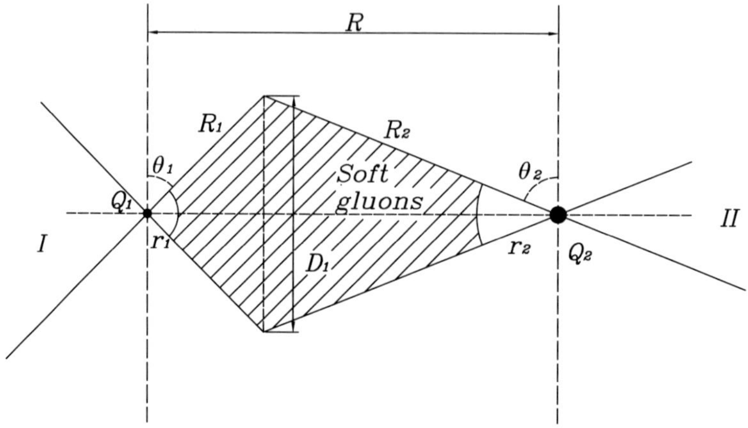

Figure 2. Vertical projection of region with the gluon condensate energy between quarks.

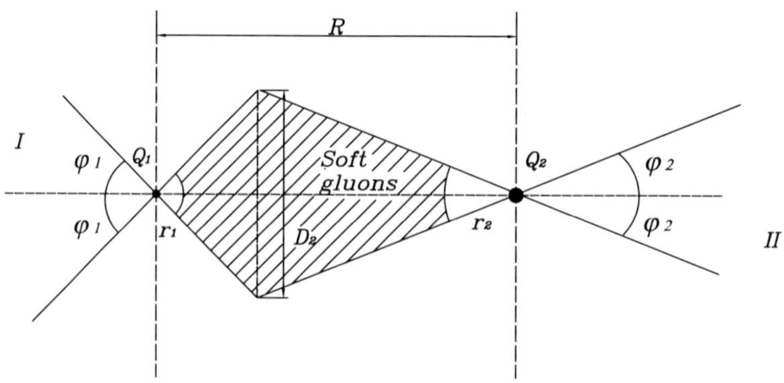

Figure 3. Horizontal projection of region with the gluon condensate energy between quarks.

schematic form of Fig. 4, where the shaded areas schematically stand for soft gluons between quarks. As a consequence, we may suppose main interaction of quarks in a baryon to be concentrated along the axes joining quarks in each pair while one may practically neglect possible interaction of any pair of quarks as a whole with third remaining quark. When approaching all three quarks the latter interaction should not increase due to asymptotical freedom so the given picture

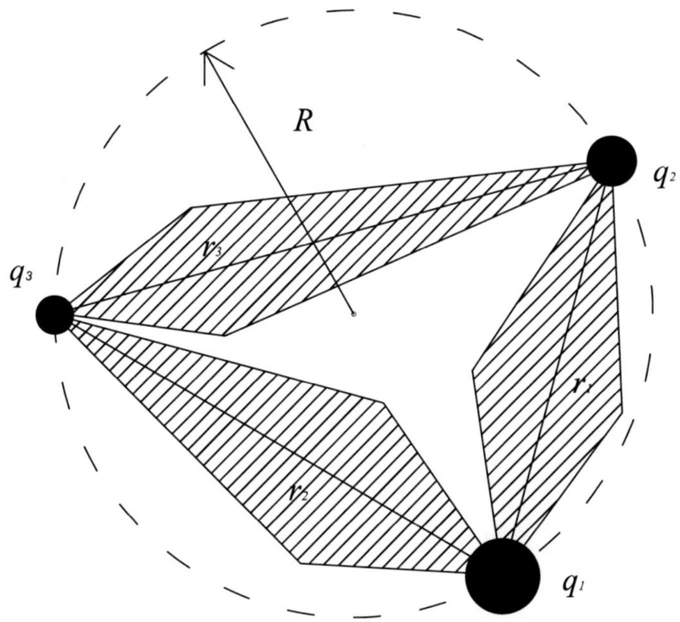

Figure 4. Schematic sketch of quark interactions in a baryon.

may in fact hold true at the scales of baryon. In what follows we consider that is the case and then we can construct the baryonic wave functions in the next way.

3.3. Baryonic Wave Functions and the Interaction Hamiltonian

In view of the above it is natural to characterize interaction of each pair of quarks in a baryon by the Hamiltonian of form (4)–(5) and, accordingly, relative motion of quarks in the pair by spherical coordinates r_i, ϑ_i, φ_i, where $i = 1$, 2, 3 correspond to pairs q_1q_2, q_1q_3, q_2q_3 respectively (see Fig. 1, 4). Then, in accordance with (4)–(5) we can introduce the Hamiltonians $H^{(i)} = (H_j^{(i)})$ with colour index $j = 1, 2, 3$. In their turn, $H^{(i)}$ will depend on $a_j^{(i)}$, $b_j^{(i)}$, $B_j^{(i)}$, i.e., the conforming parameters of solution (3) describing interaction of the ith pair of quarks [cf. (5)] while μ_0 will be the reduced mass of the corresponding quark pair. Each $H^{(i)}$ acts in Hilbert space $L_2^{12}(\mathbb{R}^3)$ which possesses the basis consisting of the eigenfunctions of $H^{(i)}$ and having the form of (6) (with the corresponding change of notations and omitting the phase factor $e^{-i\omega_j t}$). In what follows we denote such a basis by $\psi^{(i)} = (\psi_j^{(i)})$ according to (6). In virtue of smallness of baryon sizes we may put the time variables t_i for each quark pair to be equal, i.e., $t_1 = t_2 = t_3$ and then the stationary states of a baryon can be chosen in the form of tensorial product of $\psi^{(i)}$ (see Appendix C)

$$\psi = \psi^{(1)} \otimes \psi^{(2)} \otimes \psi^{(3)}, \tag{23}$$

so ψ will be a basis in the Hilbert space $\mathcal{H} = L_2^{12}(\mathbb{R}^3) \otimes L_2^{12}(\mathbb{R}^3) \otimes L_2^{12}(\mathbb{R}^3)$ and also the eigenfunctions for operator in $\mathcal{H}$ (with identical operator I)

$$H = H^{(1)} \otimes I \otimes I + I \otimes H^{(2)} \otimes I + I \otimes I \otimes H^{(3)} \tag{24}$$

with the eigenvalues $\omega = \omega_j^{(1)} + \omega_j^{(2)} + \omega_j^{(3)}$, where $\omega_j^{(i)}$ is an eigenvalue for operator $H_j^{(i)}$ (see subsection 2.1 and Appendix C), i.e.,

$$H\psi = H(\psi^{(1)} \otimes \psi^{(2)} \otimes \psi^{(3)}) = \omega\psi = (\omega_j^{(1)} + \omega_j^{(2)} + \omega_j^{(3)})\psi\,. \tag{25}$$

It is clear that operator H is just the sought interaction Hamiltonian for quarks in a baryon. As a result, the energy levels of the baryon will be given by relation $\epsilon = m_{q_1} + m_{q_2} + m_{q_3} + \omega$ with the current quark masses m_{q_k} (rest energies) of the quarks constituting the baryon and the interaction energy ω of (25).

Under the situation, the stay probability of a baryon in the volume element $dV_1dV_2dV_3$ of the configuration space with $dV_i = r_i^2 \sin\vartheta_i dr_i d\vartheta_i d\varphi_i$ will be equal to

$$dP = (\psi^{+(1)}\psi^{(1)})(\psi^{+(2)}\psi^{(2)})(\psi^{+(3)}\psi^{(3)})dV_1dV_2dV_3 \tag{26}$$

in accordance with (C.3), so

$$\int_{\mathbb{R}^3}\int_{\mathbb{R}^3}\int_{\mathbb{R}^3} dP = 1. \tag{27}$$

Having integrated dP over the angles ϑ_i, φ_i (with taking into account the properties of the eigenspinors of the Euclidean Dirac operator on two-sphere $\mathbb{S}^2$, see Appendix A), we obtain the probability of that the quarks in the baryon are at the distances r_i from each other (see Fig. 4)

$$dW = \prod_{i=1}^{3}\left(\sum_{j=1}^{3}(|F_{j1}^{(i)}|^2 + |F_{j2}^{(i)}|^2)dr_i\right). \tag{28}$$

4. Main Characteristics of Symmetrical Baryons

Obviously, we should choose a few quantities that are the most important from the physical point of view to characterize baryons under consideration and then we should evaluate the given quantities within the framework of our approach. In the circumstances let us settle on the ground state energy (mass), the root-mean-square radius and the anomalous magnetic moment. All three magnitudes are essentially nonperturbative ones, and can be calculated only by nonperturbative techniques.

4.1. Ground State Energy (Mass)

One should evidently take into account that for symmetrical baryons in question from physical point of view we have all the grounds of considering the interaction in pairs q_1q_2, q_1q_3, q_2q_3 (see Figs. 1, 4) to be the same because $q_1 = q_2 = q_3$ and, as a result, the corresponding parameters $a_j^{(i)}$, $b_j^{(i)}$ $B_j^{(i)}$ of gluonic field from solution (3) describing interaction of the ith pair of quarks should be equal: $a_j^{(i)} = a_j$, $b_j^{(i)} = b_j$, $B_j^{(i)} = B_j$. Then, in accordance with (12) and (25) we obtain the symmetrical baryon mass as

$$\mu = 3(m_q + \omega),\ \omega = \omega_j = \frac{g^2 a_j b_j}{\Lambda_j} + \frac{\alpha_j \mu_0}{|\Lambda_j|}, \tag{29}$$

or, in more explicit form,

$$\omega = \frac{g^2 a_1 b_1}{\lambda_1 - gB_1} + \mu_0 \frac{\sqrt{(\lambda_1 - gB_1)^2 - g^2 a_1^2}}{|\lambda_1 - gB_1|} =$$

$$\frac{g^2 a_2 b_2}{\lambda_2 - gB_2} + \mu_0 \frac{\sqrt{(\lambda_2 - gB_2)^2 - g^2 a_2^2}}{|\lambda_2 - gB_2|} =$$

$$\frac{g^2(a_1 + a_2)(b_1 + b_2)}{\lambda_3 + g(B_1 + B_2)} + \mu_0 \frac{\sqrt{[\lambda_3 + g(B_1 + B_2)]^2 - g^2(a_1 + a_2)^2}}{|\lambda_3 + g(B_1 + B_2)|} \tag{30}$$

with the reduced mass $\mu_0 = m_q/2$ while in what follows we take the eigenvalues of the Euclidean Dirac operator on $\mathbb{S}^2$ $\lambda_j = 1, j = 1, 2, 3$.

4.2. Baryon Radius

We can note that the root-mean-square radius of meson from (16) can be defined by the relation

$$<r>^2= -6\frac{\partial^2 F}{\partial K^2}(K=0) \tag{31}$$

with form factor $F(K)$ of (15). Obviously for symmetrical baryons it is natural to put $<r_1>^2=<r_2>^2=<r_3>^2=r_0^2$, where $<r_i>^2$ conforms to the corresponding quark pair (see Figs. 1, 4). But then simple geometric considerations give rise to the baryon radius $<R>=r_0/\sqrt{3}$ which can be defined by the relation analogous to (31)

$$<R>^2= -6\frac{\partial^2 F_E}{\partial K^2}(K=0) \tag{32}$$

if putting the baryon electric form factor $F_E(K)=\frac{1}{9}\sum\limits_{j=1}^{3}F_i(K)$, where form factor $F_i(K)$ corresponds to ith quark pair (see Figs. 1, 4), which entails the expression

$$<R>=\frac{1}{\sqrt{3}}\sqrt{\sum_{j=1}^{3}\frac{2\alpha_j^2+3\alpha_j+1}{6\beta_j^2}} \tag{33}$$

for symmetrical baryons.

4.3. Anomalous Magnetic Moment

Also it is natural to define the baryon magnetic form factor by the relation $F_M(K)=\frac{1}{3}\sum\limits_{j=1}^{3}F_i(K)$, where form factor $F_i(K)$ of (20) corresponds to ith quark pair (see Figs. 1, 4), which gives rise to the expression for the baryon anomalous magnetic moment

$$\mu_a=F_M(K=0)=\frac{2}{3}\frac{|q|}{q}\sum_{j=1}^{3}\frac{2\alpha_j+1}{6\beta_j}\cdot\frac{\mathcal{P}_j\mathcal{Q}_j}{\mathcal{P}_j^2+\mathcal{Q}_j^2},$$

$$\mathcal{P}_j=P_j(1-gb_j/\beta_j)\,,\,\mathcal{Q}_j=O_j(1+gb_j/\beta_j) \tag{34}$$

in the case of symmetrical baryons while q is electric charge of ith quark pair.

5. Numerical Estimates

5.1. Δ^{++}

At computation we took mass $\mu = 1232.0$ MeV [17] while the anomalous magnetic moment is experimentally restricted within limits $3.7 \leq \mu_a \leq 7.5$ [17] in units of nuclear magneton μ_N, so under calculation we took $\mu_a = (3.7 + 7.5)/2 = 5.6$. As for the radius $< R >$, no experimental values of it exist [17] but if taking into account that in the case of proton with mass 938 MeV we have $< R > \approx 0.875$ fm [17], then when considering the symmetrical baryons to be more compact than proton we should put, for example, 0.3 fm $\leq < R > \leq 0.7$ fm to make computation more sensible. Some numerical results are adduced in Tables 1 and 2 and each line of Table 2 corresponds to the conforming one of Table 1 while the values of μ_a are expressed in units of μ_N.

Table 1. Gauge coupling constant, reduced mass μ_0 and parameters of the confining SU(3)-gluonic field for Δ^{++}

g	μ_0 (MeV)	a_1	a_2	b_1 (GeV)	b_2 (GeV)	B_1	B_2
3.36724	1.12500	0.550635	-0.764655	-0.175434	-0.170501	1.09500	-0.7800
3.36724	1.12500	0.404093	0.203740	-0.306289	0.514750	1.3200	-0.5700
3.36724	1.12500	0.432052	-0.136707	0.318374	-0.780394	-0.8400	-0.58500

Table 2. Theoretical and experimental mass, radius and anomalous magnetic moment for Δ^{++}

Theoret. μ (MeV)	Experim. μ (MeV)	Theoret. $< R >$ (fm)	Experim. $< R >$ (fm)	Theoret. μ_a	Experim. μ_a
1232.0	$\mu = 3(m_u + \omega_j(0,0,1)) = 1232.0$	0.696633	-	5.37931	5.6
1232.0	$\mu = 3(m_u + \omega_j(0,0,1)) = 1232.0$	0.516086	-	5.77838	5.6
1232.0	$\mu = 3(m_u + \omega_j(0,0,1)) = 1232.0$	0.356102	-	6.08430	5.6

5.2. Δ^-

We here used mass $\mu = 1232.0$ MeV [17] but, in virtue of absence of the generally accepted experimental value of μ_a [17] the value $\mu_a = $ -3.68 from [28] was accepted. A number of numerical results is given by Tables 3 and 4.

Table 3. Gauge coupling constant, reduced mass μ_0 and parameters of the confining SU(3)-gluonic field for Δ^-

g	μ_0 (MeV)	a_1	a_2	b_1 (GeV)	b_2 (GeV)	B_1	B_2
3.36724	2.500	-0.708466	0.160912	-0.187350	0.332727	-0.8100	-0.1500
3.36724	2.500	0.343724	0.192053	-0.178990	0.362950	0.8100	-0.28500
3.36724	2.500	0.205870	-0.372948	-0.254934	-0.215958	0.73500	-0.37500

Table 4. Theoretical and experimental mass, radius and anomalous magnetic moment for Δ^-

Theoret. μ (MeV)	Experim. μ (MeV)	Theoret. $< R >$ (fm)	Experim. $< R >$ (fm)	Theoret. μ_a	Experim. μ_a
1232.0	$\mu = 3(m_d + \omega_j(0,0,1)) = 1232.0$	0.698664	-	-3.68497	-3.68
1232.0	$\mu = 3(m_d + \omega_j(0,0,1)) = 1232.0$	0.519021	-	-3.95427	-3.68
1232.0	$\mu = 3(m_d + \omega_j(0,0,1)) = 1232.0$	0.358006	-	-3.69648	-3.68

5.3. Ω^-

In this case the generally accepted values $\mu = 1672.45$ MeV and $\mu_a = $ -2.02 [17] were employed. Tables 5 and 6 contain some numerical results.

6. Chiral Limit and Concluding Remarks

6.1. Chiral Limit

Having obtained estimates for symmetrical baryons we can address to the chiral symmetry breaking problem in QCD whose possible resolution within

Table 5. Gauge coupling constant, reduced mass μ_0 and parameters of the confining SU(3)-gluonic field for Ω^-

g	μ_0 (MeV)	a_1	a_2	b_1 (GeV)	b_2 (GeV)	B_1	B_2
2.841363	53.7500	-0.753910	0.628741	-0.311036	-0.250568	-1.30500	1.45500
2.841363	53.7500	0.231041	0.207007	-0.194473	0.432236	0.6600	-0.28500
2.841363	53.7500	-0.406736	0.369405	-0.560202	-0.307761	-1.27500	1.15500

Table 6. Theoretical and experimental mass, radius and anomalous magnetic moment for Ω^-

Theoret. μ (MeV)	Experim. μ (MeV)	Theoret. $< R >$ (fm)	Experim. $< R >$ (fm)	Theoret. μ_a	Experim. μ_a
1672.45	$\mu = 3(m_s + \omega_j(0,0,1)) = 1672.43$	0.586809	-	-1.82490	-2.02
1672.45	$\mu = 3(m_s + \omega_j(0,0,1)) = 1672.43$	0.430573	-	-2.07908	-2.02
1672.45	$\mu = 3(m_s + \omega_j(0,0,1)) = 1672.43$	0.332305	-	-2.02136	-2.02

the framework of the above confinement mechanism has been discussed in [12]. As was mentioned in subsection 2.3, merits of case consists in that the Dirac equation in the field (3) possesses a nontrivial spectrum of bound states even for massless fermions [see relation (8′)]. As a result, mass of any meson remains nonzero in chiral limit when masses of quarks $m_q \to 0$ and meson masses will be expressed only through the parameters of the confining SU(3)-gluonic field of (3). This purely gluonic residual mass of meson should be interpreted as a gluonic contribution to the meson mass.

Physically this gives us a possible approach to the problem of chiral symmetry breaking in QCD [12]: in chirally symmetric world masses of mesons are fully determined by the confining SU(3)-gluonic field between (massless) quarks and not equal to zero. Accordingly chiral symmetry is a sufficiently rough approximation holding true only when neglecting the mentioned SU(3)-gluonic field between quarks and no additional mechanism of the spontaneous chiral symmetry breaking connected to the so-called Goldstone bosons is required. Referring for more details to [12], we can here only note that masses of symmetrical baryons have also a purely gluonic contribution and we may be interested in what part of their masses is obligatory to that contribution.

Indeed, in chiral limit $m_{q_1}, m_{q_2}, m_{q3} \to 0$ we, according to (30), obtain $(\mu)_{chiral} = 3\omega$ with

$$\omega \approx \frac{g^2 a_1 b_1}{\lambda_1 - gB_1} \approx \frac{g^2 a_2 b_2}{\lambda_2 - gB_2} \approx \frac{g^2(a_1 + a_2)(b_1 + b_2)}{\lambda_3 + g(B_1 + B_2)} \neq 0\,. \tag{35}$$

We can see that in chiral limit the baryon masses are completely determined only by the parameters a_j, b_j, B_j of SU(3)-gluonic field among quarks, i.e. by interaction among quarks, and those masses have the purely gluonic nature. Accordingly, one can use the parameters g, a_j, b_j, B_j adduced in Tables 1, 3, 5 to compute $(\mu)_{chiral}$ and then find the ratio $gl = (\mu)_{chiral}/\mu$ which in fact represents the sought gluonic contribution to the baryon masses. Then, for example, for the data of first lines of Tables 1, 3, 5 one can easy obtain $gl \approx 99.254\,\%, 98.315\,\%, 72.131\,\%$, respectively for Δ^{++}, Δ^-, Ω^-. Similar numbers hold true also for the other data from Tables 1, 3, 5.

So, even in chirally symmetric world, e.g., the symmetric baryons would have nonzero masses that would be determined only by SU(3)-gluonic interaction among massless quarks, i.e. those masses would have a purely gluonic nature. Moreover, since gluons are verily relativistic particles then the most part of masses for baryons under discussion is conditioned by relativistic effects, as is seen from the above estimates. Further discussion of the proposed chiral symmetry breaking mechanism can be found in [12, 13].

6.2. Concluding Remarks

It should be noted that modern baryon spectroscopy in theoretical aspects can not be considered satisfactory enough: the main theoretical tools for describing dynamics of quarks in baryons here are miscellaneous potential models (see, e.g., early and recent reviews [29, 30]). Quark interaction at such a description is modelled mainly by either oscillator potential or the confining potential of form $a/r + br$. But, as follows from the uniqueness theorem of Section 2 (see also Appendix D), such potentials cannot describe any gluon configuration between quarks. It would be possible if the mentioned potentials were solutions of Yang-Mills equations directly derived from QCD-Lagrangian since, from the QCD-point of view, any gluonic field should be a solution of Yang-Mills equations (as well as any electromagnetic field is by definition always a solution of Maxwell equations). According to the above uniqueness theorem neither oscillator potential nor the confining potential of form $a/r + br$ are not solutions of Yang-Mills equations. In particular, Coulomb and linear parts belong to different parts of the YM-potentials (see (3)), accordingly, to the colour electric and colour magnetic parts so they cannot be united in one component of form $a/r + br$. So, we draw the conclusion (mentioned as far back as in Refs. [4, 5] and elaborated more in detail in Ref. [7]) that the potential approach seems to be inconsistent.

Another matter is our confinement mechanism proposed in [1, 2, 3] which gives new possibilities from the first principles of QCD immediately appealing to the quark and gluonic degrees of freedom. Indeed, the words *quark and gluonic degrees of freedom* make exact sense: gluons come forward in the form of bosonic condensate described by parameters a_j, b_j, B_j from the unique exact solution (3) of the Yang-Mills equations while quarks are represented by their current masses m_q.

Consequently, the given mechanism is based on the unique family of compatible nonperturbative solutions for the Dirac-Yang-Mills system directly derived from QCD-Lagrangian and, as a result, the approach is itself nonperturbative, relativistic from the outset, admits self-consistent nonrelativistic limit and may be employed for any meson (quarkonium). Applications of the approach in Refs. [4, 5, 6, 7, 8, 9, 10, 11, 12, 13] allow one to speak about the fact that the confinement mechanism elaborated in [1, 2, 3] gives new possibilities for considering many old problems of hadronic (meson) physics (such as nonperturbative computation of decay constants, masses and radii of mesons, chiral symmetry breaking, the origin of the so-called scale Λ_{QCD} [47] and so forth) from the first principles of QCD immediately appealing to the quark and gluonic degrees of freedom.

At the same time, as the results of the present Chapter show, this mechanism may be extended over baryons in a reasonable way. We hope further study of baryons with the help of the given approach to shed new light on the internal structure of baryons.

7. Problem of Masses in Particle Physics

7.1. Preliminaries

As is known [17], the generally accepted standard model with one Higgs doublet asserts that the masses of fundamental fermions (quarks and leptons) are acquired through the Higgs mechanism so for their masses m_i we obtain (without taking mixings into account) $m_i = f_i v/\sqrt{2}$, where the vacuum Higgs condensate $v \approx 246$ GeV and i stands for quark and lepton flavours. But little is known about the coupling constants f_i and much may be elucidated only with discovering Higgs bosons. The same holds true for the gauge bosons $W^{\pm}$, Z where masses $m_W = ev/(2\sin\theta_W)$, $m_Z = ev/(\sin 2\theta_W)$ with the so-called weak angle θ_W so that $\sin^2\theta_W \approx 0.23$ and e is the elementary electric charge. If taking into account that the mass of Higgs boson $m_H = \lambda v$ with a self-interaction constant λ then it is clear that masses of all the abovementioned particles are proportional to m_H, and, consequently, the discovery of Higgs boson will not completely resolve the puzzle of origin of masses in particle physics – the question will remain where the mass m_H comes from not speaking already about the nature of the above miscellaneous constants f_i and λ.

At present, to our mind, one can single out two most promising approaches to a possible resolution for the mentioned problems: technicolour theories and preon models. Under the circumstances let us shortly outline how both these directions might be estimated from the point of view of our confinement mechanism and the chiral symmetry breaking one based on the latter and discussed above and in [12].

7.2. Technicolour Theories

Referring for more details concerning those models to both early references [31, 32, 33] and modern status of them (see, e.g., [34]) let us note the following. The main idea of acquiring masses, e.g., for $W^{\pm}$ and Z bosons, consists in that a new set of the so-called techniquarks is postulated at the energy scale of order 1 TeV which interact with each other through the technigluons and it makes the massless techipions exist as Goldstone bosons. The latter give masses to $W^{\pm}$ and Z after spontaneous symmetry breaking. It should be noted, however, those massless technipions appear as a result of violating chiral symmetry connected with technicolour QCD on the analogy with chiral symmetry breaking in usual QCD. But, as we have discussed in [12] and in Section 6, the hypothetical mechanism for chiral symmetry breaking in standard QCD with appearance of Goldstone bosons (pions) seems to fail because of pions can never be massless inasmuch as they have nonzero masses even in chirally symmetric world due to gluons. The same will also perfectly hold true for technipions which would always have nonzero masses due to technigluons since technicolour QCD should manifest the confinement mechanism similar to our one in usual QCD. Therefore, technicolour theories look rather doubtful from the point of view of our confinement mechanism.

7.3. Preon Models

Another cardinal approach to the problem of masses is connected with the preon models (see, e.g., [35] and references therein). Under this approch quarks, leptons and gauge vector bosons are suggested to be composed of stable spin-1/2 preons, for example, existing in three flavours and being combined according to simple rules. The main theoretical objection to preon theories is the mass paradox which arises by virtue of the Heisenberg's uncertainty principle. Scattering experiments have shown [16] that quarks and leptons are point-like up to the scales of order 10^{-3} fm which corresponds to a preon mass of order 197 GeV (due to the uncertainty principle) if the preon is confined to a box of such a size, i.e. its mass will approximately be 0.4×10^5 times greater than, e.g., that of d-quark. Thus, the preon models are faced with a mass paradox: how could quarks or electrons be made of smaller particles that would have masses of many orders of magnitude greater than the fundamental fermion masses? The paradox might be resolved by the rather dubious postulate about a large binding force between preons cancelling their mass-energies. Our confinement mechanism points out the more physically acceptable way of overcoming these obstacles. If the interaction among preons is decribed by a QCD-like theory based on, e.g., SU(N)-group with $N \geq 2$ then, according to our results [2, 3], such theories should also manifest confinement to generate masses decribed by relations similar to (8) and $(8')$. This signifies that preons might possess small masses or be just massless and, as a result, mass paradox would be removed.

Among fundamental fermions one should pay special attention to electron. Within the framework of preon models the most natural composition for it should be the one of three preons [35], so electron might be described by triplet (β, β, δ), where preons β and δ carry electric charges $-2e/3$, $e/3$, respectively, with elementary charge e [35]. If supposing that interaction among them is desribed by the gauge group SU(3) through pre-gluons [35] then we formally come to the same scheme which was under exploration for baryons in the given chapter. Under the circumstances we can find the parameters in solution (3) which will correspond to the observed mass, some radius (of order of the classical electron radius $r_e \sim 10^{15}$ m) and (anomalous) magnetic moment for electron. In those calculations one can consider the mentioned preons to be massless, as is said above. We hope to study this interesting task elsewhere.

8. Remarks about the Yang-Mills Millennium Problem

Our approach to the quark confinement mechanism is closely connected with the so-called Yang-Mills Millennium problem (see [36] for more details). Its formulation looks as follows:

Yang-Mills Existence and Mass Gap: *Prove that for any compact simple gauge group G, quantum Yang-Mills theory on $\mathbb{R}^4$ exists and has a mass gap* $\Delta > 0$.

It is not complicated to see that our results on the confinement mechanism [in particular, the solution (3)] to a certain degree clear the way to obtain a proof of the above problem. So long as group SU(3) belongs to the class of groups G from the formulation of problem then let us demonstrate possible opportunities with an example of the SU(3) gauge group and then make some remarks concerning the common case.

Let us note the following. From mathematical point of view the Yang-Mills field are the *connections* in vector (or, which is equivalent, in principal) bundles over some manifold M. When M is the Minkowski space the situation is simplified so long as all the bundles over M with topology $\mathbb{R}^4$ are trivial. When quantising Yang-Mills field we obtain, from physical point of view, quanta of that field, e.g., photons or gluons, which realize an interaction among the corresponding (matter) particles able to interchange by those quanta. But from mathematical point of view the mentioned particles are described by the *cross-sections* of vector bundles over the given spacetime, in particular, over $\mathbb{R}^4$. As a result, mathematically, connections do not live separately from cross-sections while, physically, the sources of any quanta are the matter particles.

Under the circumstances when discussing the above millennium problem we should take into account also the matter fields rather than restrict ourselves to the pure Yang-Mills theory and further we should pass on to the pure Yang-Mills case by some limiting procedure. Then, returning to group SU(3), we have seen that there exists the exact solution (3) of the Yang-Mills equations which is unique in a certain sense (see Appendix D) and spectrum of bound states in such a field is given by the relation (8). The latter formula cannot be obtained without including spinor fields as cross-sections of the corresponding bundle into consideration.

We now especially are interested in the limiting case of (8), namely in the relation $(8')$. In it all the information about matter (spinor) fields vanished since that information was related only with mass parameters in the conforming Dirac equation. As a consequence, the relation $(8')$ should be interpreted as the pure gluonic energy. Obviously, the latter only depends on quantities a_j, b_j, B_j created by group SU(3) and is quantized in virtue of availability of the nonnegative integer numbers n_j. Also from considerations of Section 2 it is clear that the next relation should be fulfilled

$$(\omega_1)_{\text{chiral}} = (\omega_2)_{\text{chiral}} = (\omega_3)_{\text{chiral}}, \tag{36}$$

which gives an additional number of restrictions on the parameters a_j, b_j, B_j. Further we can see that there is a mass gap Δ defined at $n_j = 0$, $\lambda_j = 0$ from the relation

$$\Delta = \frac{a_1 b_1}{\pm 1 - g B_1} = \frac{a_2 b_2}{\pm 1 - g B_2} = \frac{(a_1 + a_2)(b_1 + b_2)}{\pm 1 + g(B_1 + B_2)}. \tag{37}$$

Of course, Δ can be positive or negative depending from a choice of parameters a_j, b_j, B_j but parameters a_j, b_j should not be equal to 0. I. e., we obtained a quantum theory for SU(3)-group asserting that any reasonable energy connected with its quanta (gluons) is quantized and has a mass gap. As an application of those properties we have already discussed the chiral symmetry breaking mechanism in Sections 2 and 6 and in Refs. [12, 13].

Passing on to the general case, we can note that (see Appendix D) the obtained results may be extended over all semisimple compact Lie groups since for them the corresponding Lie algebras possess just the only Cartan subalgebra. Also we can talk about the compact non-semisimple groups, for example, U(N). In the latter case additionally to Cartan subalgebra we have centrum consisting from the matrices of the form αI_N (I_N is the unit matrix $N \times N$) with arbitrary constant α. For all those groups there exists their own uniqueness theorem, as follows from considerations in Appendix D, one can evaluate spectrum of bound states (examples for groups SU(2) and SU(4) can be found in [3]) and chiral limit that entails results about quantizing energy of the corresponding quanta and gives a mass gap for all those groups. To get the formulas in an explicit form one should take a concrete realization of the conforming Lie algebras.

The most relevant physical cases are of course U(1)- and SU(3)-ones (QED and QCD). In particular, the U(1)-case allows us to build the classical model of confinement (see Section 2 and Ref. [46]).

Conclusion

The main idea of quark confinement may be borrowed from classical electrodynamics. Indeed, let us recall the well-known case of motion of a charged particle in the homogeneous magnetic field (see, e.g., Ref. [25]). In the latter case the particle moves along helical curve with lead of helix $h = 2\pi mv\cos\alpha/(qH\sqrt{1-v^2})$ and radius $R = mv\sin\alpha/(qB\sqrt{1-v^2})$, where α is an angle between vectors of the particle velocity $\mathbf{v}$ and magnetic induction $\mathbf{B}$, q is particle charge, m is particle mass. As a consequence, the homogeneous magnetic field does not give rise to the full confinement of the particle since the latter may go to infinity along the helical curve. The situation is not changed at quantum level as well: there exist no bound states in the homogeneous magnetic field [27]. But if estimating module of B at $R \sim 10^{-15}\,\mathrm{m} = 1$ fm for electron then B will be of order 10^{23} T. I. e., if considering that quarks are confined by a colour magnetic field that they themselves create then one needs a colour magnetic field between quarks with B of such an order and that field should not allow any quark to go to infinity. It is clear this field should be a solution of the Yang-Mills equations. Really, if speaking about Minkowski spacetime then searching for classical solutions of the Yang-Mills equations makes sense because at large distances between quarks the latter are surrounded with a huge number of gluons that are emitted by both quarks and gluons themselves. Under this situation it is quite plausible that confinement of quarks arises due to certain properties of such gluonic clouds while the latter should be described just by *classical* solutions of the Yang-Mills equations.

As was shown in Refs. [1, 2, 3], the necessary solution has the form (3) for group SU(3) and is unique in a certain sense. This result allowed us to propose a quark confinement mechanism which was successfully applied to meson spectroscopy. Indeed, as follows from (16) at $|b_j| \to \infty$ we have $< r > \sim \sqrt{\sum_{j=1}^{3} \frac{1}{(g|b_j|)^2}}$, so in the strong magnetic colour field when $|b_j| \to \infty$, $< r > \to 0$, while the meson wave functions of (6) and (9) behave as $\Psi_j \sim e^{-g|b_j|r}$, i. e., just the magnetic colour field of (3) provides two quarks with confinement. This situation also holds true at classical level [46].

The given Chapter extends this mechanism over baryons and further study of that approach will be connected with analysis of concrete baryons with its help.

Appendix A

We here represent some results about eigenspinors of the Euclidean Dirac operator on two-sphere $\mathbb{S}^2$ employed in the main part of the Chapter.

When separating variables in the Dirac equation (4) there naturally arises the Euclidean Dirac operator $\mathcal{D}_0$ on the unit two-dimensional sphere $\mathbb{S}^2$ and we should know its eigenvalues with the corresponding eigenspinors. Such a problem also arises in the black hole theory while describing the so-called twisted spinors on Schwarzschild and Reissner-Nordström black holes and it was analysed in Refs. [3, 37, 38], so we can use the results obtained therein for our aims. Let us adduce the necessary relations.

The eigenvalue equation for corresponding spinors Φ may look as follows

$$\mathcal{D}_0\Phi = \lambda\Phi. \tag{A.1}$$

As was discussed in Refs. [37, 38] the natural form of $\mathcal{D}_0$ (arising within applications) in local coordinates ϑ, φ on the unit sphere $\mathbb{S}^2$ looks as

$$\mathcal{D}_0 = -i\sigma_1\left[i\sigma_2\partial_\vartheta + i\sigma_3\frac{1}{\sin\vartheta}\left(\partial_\varphi - \frac{1}{2}\sigma_2\sigma_3\cos\vartheta\right)\right] =$$

$$\sigma_1\sigma_2\partial_\vartheta + \frac{1}{\sin\vartheta}\sigma_1\sigma_3\partial_\varphi + \frac{\cot\vartheta}{2}\sigma_1\sigma_2 \tag{A.2}$$

with the ordinary Pauli matrices

$$\sigma_1 = \begin{pmatrix} 0 & 1 \\ 1 & 0 \end{pmatrix}, \sigma_2 = \begin{pmatrix} 0 & -i \\ i & 0 \end{pmatrix}, \sigma_3 = \begin{pmatrix} 1 & 0 \\ 0 & -1 \end{pmatrix},$$

so that $\sigma_1\mathcal{D}_0 = -\mathcal{D}_0\sigma_1$.

The equation (A.1) was explored in Refs. [37, 38]. Spectrum of D_0 consists of the numbers $\lambda = \pm(l+1)$ with multiplicity $2(l+1)$ of each one, where $l = 0, 1, 2, ...$ Let us introduce the number m such that $-l \leq m \leq l+1$ and the corresponding number $m' = m - 1/2$ so $|m'| \leq l + 1/2$. Then the conforming eigenspinors of operator $\mathcal{D}_0$ are

$$\Phi = \begin{pmatrix} \Phi_1 \\ \Phi_2 \end{pmatrix} = \Phi_{\mp\lambda} = \frac{C}{2} \begin{pmatrix} P^k_{m'-1/2} \pm P^k_{m'1/2} \\ P^k_{m'-1/2} \mp P^k_{m'1/2} \end{pmatrix} e^{-im'\varphi} \tag{A.3}$$

with the coefficient $C = \sqrt{\frac{l+1}{2\pi}}$ and $k = l + 1/2$. These spinors form an orthonormal basis in $L_2^2(\mathbb{S}^2)$ and are subject to the normalization condition

$$\int_{\mathbb{S}^2} \Phi^\dagger \Phi d\Omega = \int_0^\pi \int_0^{2\pi} (|\Phi_1|^2 + |\Phi_2|^2) \sin\vartheta d\vartheta d\varphi = 1\,. \tag{A.4}$$

Further, owing to the relation $\sigma_1 \mathcal{D}_0 = -\mathcal{D}_0 \sigma_1$ we, obviously, have

$$\sigma_1 \Phi_{\mp\lambda} = \Phi_{\pm\lambda}\,. \tag{A.5}$$

As to functions $P^k_{m'n'}(\cos\vartheta) \equiv P^k_{m',n'}(\cos\vartheta)$ then they can be chosen by miscellaneous ways, for instance, as follows (see, e. g., Ref. [39])

$$P^k_{m'n'}(\cos\vartheta) =$$

$$i^{-m'-n'} \sqrt{\frac{(k-m')!(k-n')!}{(k+m')!(k+n')!}} \left(\frac{1+\cos\vartheta}{1-\cos\vartheta}\right)^{\frac{m'+n'}{2}} \times$$

$$\times \sum_{j=\max(m',n')}^{k} \frac{(k+j)!\,i^{2j}}{(k-j)!(j-m')!(j-n')!} \left(\frac{1-\cos\vartheta}{2}\right)^j \tag{A.6}$$

with the orthogonality relation at m', n' fixed

$$\int_0^\pi P^{*k}_{m'n'}(\cos\vartheta) P^{k'}_{m'n'}(\cos\vartheta) \sin\vartheta d\vartheta = \frac{2}{2k+1}\delta_{kk'}\,. \tag{A.7}$$

It should be noted that square of $\mathcal{D}_0$ is

$$\mathcal{D}_0^2 = -\Delta_{\mathbb{S}^2} I_2 + \sigma_2\sigma_3 \frac{\cos\vartheta}{\sin^2\vartheta}\partial_\varphi + \frac{1}{4\sin^2\vartheta} + \frac{1}{4}\,, \tag{A.8}$$

while laplacian on the unit sphere is

$$\Delta_{\mathbb{S}^2} = \frac{1}{\sin\vartheta}\partial_\vartheta \sin\vartheta\partial_\vartheta + \frac{1}{\sin^2\vartheta}\partial^2_\varphi = \partial^2_\vartheta + \cot\vartheta\partial_\vartheta + \frac{1}{\sin^2\vartheta}\partial^2_\varphi\,, \qquad (A.9)$$

so the relation (A.8) is a particular case of the so-called Weitzenböck-Lichnerowicz formulas (see Refs. [40]). Then from (A.1) it follows $\mathcal{D}_0^2\Phi = \lambda^2\Phi$ and, when using the ansatz $\Phi = P(\vartheta)e^{-im'\varphi} = \begin{pmatrix} P_1 \\ P_2 \end{pmatrix} e^{-im'\varphi}$, $P_{1,2} = P_{1,2}(\vartheta)$, the equation $\mathcal{D}_0^2\Phi = \lambda^2\Phi$ turns into

$$\left(-\partial^2_\vartheta - \cot\vartheta\partial_\vartheta + \frac{m'^2+\frac{1}{4}}{\sin^2\vartheta} + \frac{m'\cos\vartheta}{\sin^2\vartheta}\sigma_1\right)P =$$

$$\left(\lambda^2 - \frac{1}{4}\right)P\,, \qquad (A.10)$$

wherefrom all the above results concerning spectrum of $\mathcal{D}_0$ can be derived [37, 38].

When calculating the functions $P^k_{m'n'}(\cos\vartheta)$ directly, to our mind, it is the most convenient to use the integral expression [39]

$$P^k_{m'n'}(\cos\vartheta) = \frac{1}{2\pi}\sqrt{\frac{(k-m')!(k+m')!}{(k-n')!(k+n')!}}$$

$$\int_0^{2\pi}\left(e^{i\varphi/2}\cos\frac{\vartheta}{2} + ie^{-i\varphi/2}\sin\frac{\vartheta}{2}\right)^{k-n'} \times$$

$$\left(ie^{i\varphi/2}\sin\frac{\vartheta}{2} + e^{-i\varphi/2}\cos\frac{\vartheta}{2}\right)^{k+n'} e^{im'\varphi}d\varphi \qquad (A.11)$$

and the symmetry relations ($z = \cos\vartheta$)

$$P^k_{m'n'}(z) = P^k_{n'm'}(z),\; P^k_{m',-n'}(z) = P^k_{-m',n'}(z),$$

$$P^k_{m'n'}(z) = P^k_{-m',-n'}(z)\,,$$

$$P^k_{m'n'}(-z) = i^{2k-2m'-2n'}P^k_{m',-n'}(z)\,. \qquad (A.12)$$

In particular

$$P^k_{kk}(z) = \cos^{2k}(\vartheta/2),\, P^k_{k,-k}(z) = i^{2k}\sin^{2k}(\vartheta/2),$$

$$P^k_{k0}(z) = \frac{i^k\sqrt{(2k)!}}{2^k k!}\sin^k\vartheta\,,$$

$$P^k_{kn'}(z) = i^{k-n'}\sqrt{\frac{(2k)!}{(k-n')!(k+n')!}}\times$$

$$\sin^{k-n'}(\vartheta/2)\cos^{k+n'}(\vartheta/2)\,. \tag{A.13}$$

Eigenspinors with $\lambda = \pm 1,\ \pm 2$

If $\lambda = \pm(l+1) = \pm 1$ then $l = 0$ and from (A.3) it follows that $k = l + 1/2 = 1/2$, $|m'| \leq 1/2$ and we need the functions $P^{1/2}_{m',\pm 1/2}$ that are easily evaluated with the help of (A.11)–(A.13) so the eigenspinors for $\lambda = -1$ are

$$\Phi = \frac{C}{2}\begin{pmatrix}\cos\frac{\vartheta}{2} + i\sin\frac{\vartheta}{2}\\ \cos\frac{\vartheta}{2} - i\sin\frac{\vartheta}{2}\end{pmatrix}e^{i\varphi/2},$$

$$\Phi = \frac{C}{2}\begin{pmatrix}\cos\frac{\vartheta}{2} + i\sin\frac{\vartheta}{2}\\ -\cos\frac{\vartheta}{2} + i\sin\frac{\vartheta}{2}\end{pmatrix}e^{-i\varphi/2}, \tag{A.14}$$

while for $\lambda = 1$ the conforming spinors are

$$\Phi = \frac{C}{2}\begin{pmatrix}\cos\frac{\vartheta}{2} - i\sin\frac{\vartheta}{2}\\ \cos\frac{\vartheta}{2} + i\sin\frac{\vartheta}{2}\end{pmatrix}e^{i\varphi/2},$$

$$\Phi = \frac{C}{2}\begin{pmatrix}-\cos\frac{\vartheta}{2} + i\sin\frac{\vartheta}{2}\\ \cos\frac{\vartheta}{2} + i\sin\frac{\vartheta}{2}\end{pmatrix}e^{-i\varphi/2} \tag{A.15}$$

with the coefficient $C = \sqrt{1/(2\pi)}$.

It is clear that (A.14)–(A.15) can be rewritten in the form

$$\lambda = -1 : \Phi = \frac{C}{2}\begin{pmatrix}e^{i\frac{\vartheta}{2}}\\ e^{-i\frac{\vartheta}{2}}\end{pmatrix}e^{i\varphi/2},$$

or

$$\Phi = \frac{C}{2}\begin{pmatrix}e^{i\frac{\vartheta}{2}}\\ -e^{-i\frac{\vartheta}{2}}\end{pmatrix}e^{-i\varphi/2},$$

$$\lambda = 1 : \Phi = \frac{C}{2}\begin{pmatrix}e^{-i\frac{\vartheta}{2}}\\ e^{i\frac{\vartheta}{2}}\end{pmatrix}e^{i\varphi/2},$$

or

$$\Phi = \frac{C}{2}\begin{pmatrix}-e^{-i\frac{\vartheta}{2}}\\ e^{i\frac{\vartheta}{2}}\end{pmatrix}e^{-i\varphi/2}, \tag{A.16}$$

so the relation (A.5) is easily verified at $\lambda = \pm 1$.

In studying vector mesons and excited states of heavy quarkonia eigenspinors with $\lambda = \pm 2$ may also be useful. Then $k = l + 1/2 = 3/2$, $|m'| \leq 3/2$ and we need the functions $P^{3/2}_{m',\pm 1/2}$ that can be evaluated with the help of (A.11)–(A.13). Computation gives rise to

$$P^{3/2}_{3/2,-1/2} = -\frac{\sqrt{3}}{2}\sin\vartheta\sin\frac{\vartheta}{2} = P^{3/2}_{-3/2,1/2},$$

$$P^{3/2}_{3/2,1/2} = i\frac{\sqrt{3}}{2}\sin\vartheta\cos\frac{\vartheta}{2} = P^{3/2}_{-3/2,-1/2},$$

$$P^{3/2}_{1/2,-1/2} = -\frac{i}{4}\left(\sin\frac{\vartheta}{2} - 3\sin\frac{3}{2}\vartheta\right) = P^{3/2}_{-1/2,1/2},$$

$$P^{3/2}_{1/2,1/2} = \frac{1}{4}\left(\cos\frac{\vartheta}{2} + 3\cos\frac{3}{2}\vartheta\right) = P^{3/2}_{-1/2,-1/2}, \qquad \text{(A.17)}$$

and according to (A.3) this entails eigenspinors with $\lambda = 2$ in the form

$$\frac{C}{2}i\frac{\sqrt{3}}{2}\sin\vartheta\begin{pmatrix} e^{-i\frac{\vartheta}{2}} \\ e^{i\frac{\vartheta}{2}}\end{pmatrix}e^{i3\varphi/2},\ \frac{C}{8}\begin{pmatrix} 3e^{-i\frac{3\vartheta}{2}} + e^{i\frac{\vartheta}{2}} \\ 3e^{i\frac{3\vartheta}{2}} + e^{-i\frac{\vartheta}{2}}\end{pmatrix}e^{i\varphi/2},$$

$$\frac{C}{8}\begin{pmatrix} -3e^{-i\frac{3\vartheta}{2}} - e^{i\frac{\vartheta}{2}} \\ 3e^{i\frac{3\vartheta}{2}} + e^{-i\frac{\vartheta}{2}}\end{pmatrix}e^{-i\varphi/2},\ \frac{C}{2}i\frac{\sqrt{3}}{2}\sin\vartheta\begin{pmatrix} -e^{-i\frac{\vartheta}{2}} \\ e^{i\frac{\vartheta}{2}}\end{pmatrix}e^{-i3\varphi/2} \qquad \text{(A.18)}$$

with $C = 1/\sqrt{\pi}$, while eigenspinors with $\lambda = -2$ are obtained in accordance with relation (A.5).

Appendix B

We here adduce the explicit form for the radial parts of meson wave functions from (6). At $n_j = 0$ they are given by

$$F_{j1} = C_j P_j r^{\alpha_j} e^{-\beta_j r}\left(1 - \frac{Y_j}{Z_j}\right),$$

$$F_{j2} = iC_j Q_j r^{\alpha_j} e^{-\beta_j r}\left(1 + \frac{Y_j}{Z_j}\right), \qquad \text{(B.1)}$$

while at $n_j > 0$ they are given by

$$F_{j1} = C_j P_j r^{\alpha_j} e^{-\beta_j r} \times$$
$$\left[\left(1 - \frac{Y_j}{Z_j}\right) L^{2\alpha_j}_{n_j}(r_j) + \frac{P_j Q_j}{Z_j} r_j L^{2\alpha_j+1}_{n_j-1}(r_j)\right],$$
$$F_{j2} = iC_j Q_j r^{\alpha_j} e^{-\beta_j r} \times$$
$$\left[\left(1 + \frac{Y_j}{Z_j}\right) L^{2\alpha_j}_{n_j}(r_j) - \frac{P_j Q_j}{Z_j} r_j L^{2\alpha_j+1}_{n_j-1}(r_j)\right] \tag{B.2}$$

with the Laguerre polynomials $L^\rho_n(r_j)$, $r_j = 2\beta_j r$, $\beta_j = \sqrt{\mu_0^2 - \omega_j^2 + g^2 b_j^2}$ at $j = 1, 2, 3$ with $b_3 = -(b_1 + b_2)$, $P_j = gb_j + \beta_j$, $Q_j = \mu_0 - \omega_j$, $Y_j = P_j Q_j \alpha_j + (P_j^2 - Q_j^2) g a_j/2$, $Z_j = P_j Q_j \Lambda_j + (P_j^2 + Q_j^2) g a_j/2$ with $a_3 = -(a_1 + a_2)$, $\Lambda_j = \lambda_j - gB_j$ with $B_3 = -(B_1 + B_2)$, $\alpha_j = \sqrt{\Lambda_j^2 - g^2 a_j^2}$, while $\lambda_j = \pm(l_j + 1)$ are the eigenvalues of Euclidean Dirac operator $\mathcal{D}_0$ on unit two-sphere with $l_j = 0, 1, 2, ...$ (see Appendix A) and quantum numbers $n_j = 0, 1, 2, ...$ are defined by the relations

$$n_j = \frac{gb_j Z_j - \beta_j Y_j}{\beta_j P_j Q_j}, \tag{B.3}$$

which entails the spectrum (8). Further, C_j of (B.1)–(B.2) should be determined from the normalization condition

$$\int_0^\infty (|F_{j1}|^2 + |F_{j2}|^2) dr = \frac{1}{3}. \tag{B.4}$$

As a consequence, we shall gain that in (4) $\Psi_j \in L_2^4(\mathbb{R}^3)$ at any $t \in \mathbb{R}$ and, accordingly, $\Psi = (\Psi_1, \Psi_2, \Psi_3)$ may describe relativistic bound states in the field (3) with the energy spectrum (8). As is clear from (B.3) at $n_j = 0$ we have $gb_j/\beta_j = Y_j/Z_j$ so the radial parts of (B.1) can be rewritten as

$$F_{j1} = C_j P_j r^{\alpha_j} e^{-\beta_j r} \left(1 - \frac{gb_j}{\beta_j}\right),$$
$$F_{j2} = iC_j Q_j r^{\alpha_j} e^{-\beta_j r} \left(1 + \frac{gb_j}{\beta_j}\right). \tag{B.5}$$

More details can be found in Refs. [1, 3].

Appendix C

We here recall some facts about tensorial products of Hilbert spaces necessary in the main part of the Chapter. For more information see, e.g., Ref. [39] and references therein.

For two Hilbert spaces $\mathcal{H}_1$ and $\mathcal{H}_2$, by definition, their tensorial product $\mathcal{H}_1 \otimes \mathcal{H}_2$ consists from every possible linear combinations of elements of the form $\vec{x}_1 \otimes \vec{x}_2$, where $\vec{x}_1 \in \mathcal{H}_1$, $\vec{x}_2 \in \mathcal{H}_2$. The obvious relations take place

$$(\alpha\vec{x}_1 + \beta\vec{x}_2) \otimes \vec{y} = \alpha(\vec{x}_1 \otimes \vec{y}) + \beta(\vec{x}_2 \otimes \vec{y}) , \tag{C.1}$$

$$x \otimes (\alpha\vec{y}_1 + \beta\vec{y}_2) = \alpha(\vec{x} \otimes \vec{y}_1) + \beta(\vec{x} \otimes \vec{y}_2) \tag{C.2}$$

with arbitrary complex numbers α, β.

Scalar product in $\mathcal{H}_1 \otimes \mathcal{H}_2$ is defined by the equality

$$(\vec{x}_1 \otimes \vec{y}_1, \vec{x}_2 \otimes \vec{y}_2) = (x_1, x_2)_1 (y_1, y_2)_2 , \tag{C.3}$$

where $(x_1, x_2)_1$, $(y_1, y_2)_2$ are the scalar products, respectively, in $\mathcal{H}_1$ and $\mathcal{H}_2$. Under the situation, if $\{\vec{e}_i\}$ is an orthonormal basis in $\mathcal{H}_1$ and $\{\vec{f}_j\}$ is an orthonormal basis in $\mathcal{H}_2$ then $\vec{h}_{ij} = \vec{e}_i \otimes \vec{f}_j$ is an orthonormal basis in $\mathcal{H}_1 \otimes \mathcal{H}_2$.

Further, if A is a linear operator in $\mathcal{H}_1$ and B is a linear operator in $\mathcal{H}_2$ then tensorial (Kronecker) product $A \otimes B$ acts in $\mathcal{H}_1 \otimes \mathcal{H}_2$ and is defined by relation $(A \otimes B)(\vec{x}_1 \otimes \vec{x}_2) = A\vec{x}_1 \otimes B\vec{x}_2$ with the properties

$$A \otimes (\alpha B_1 + \beta B_2) = \alpha A \otimes B_1 + \beta A \otimes B_2 , \tag{C.4}$$

$$(\alpha A_1 + \beta A_2) \otimes B = \alpha A_1 \otimes B + \beta A_2 \otimes B , \tag{C.5}$$

$$A_1 A_2 \otimes B_1 B_2 = (A_1 \otimes B_1)(A_2 \otimes B_2) . \tag{C.6}$$

In a similar way, the Kronecker sum of A and B is defined by $A \oplus B = A \otimes I + I \otimes B$ with identical operator I. Under the circumstances, let λ_i be the eigenvalues A and μ_j be those of B (listed according to multiplicity). Then the eigenvalues of $A \otimes B$ are $\lambda_i\mu_j$ while those of $A \oplus B$ will be $\lambda_i + \mu_j$.

All the above is directly generalized to the arbitrary number of Hilbert spaces. For example, the operator $A_1 \otimes I \otimes I + I \otimes A_2 \otimes I + I \otimes I \otimes A_3$ has the eigenvalues $\lambda_i + \mu_j + \nu_k$ in space $\mathcal{H}_1 \otimes \mathcal{H}_2 \otimes \mathcal{H}_3$ if λ_i, μ_j, ν_k are the eigenvalues of A_1, A_2 and A_3, respectively, in $\mathcal{H}_1$, $\mathcal{H}_2$, $\mathcal{H}_3$.

Appendix D

The facts adduced here have been obained in Refs. [2, 3] and we concisely give them only for completeness of discussion in Section 2.

To specify the question, let us note that in general the Yang-Mills equations on a manifold M can be written as

$$d * F = g(*F \wedge A - A \wedge *F)\,, \tag{D.1}$$

where a gluonic field $A = A_\mu dx^\mu = A^a_\mu \lambda_a dx^\mu$ [λ_a are the known Gell-Mann matrices, $\mu = t, r, \vartheta, \varphi$ (in the case of spherical coordinates), $a = 1, ..., 8$], the curvature matrix (field strentgh) $F = dA + gA \wedge A = F^a_{\mu\nu}\lambda_a dx^\mu \wedge dx^\nu$ with exterior differential d and the Cartan's (exterior) product $\wedge$, while $*$ means the Hodge star operator conforming to a metric on manifold under consideration, g is a gauge coupling constant.

The most important case of M is Minkowski spacetime and we are interested in the confining solutions A of the SU(3)-Yang-Mills equations. The confining solutions were defined in Ref. [1] as the spherically symmetric solutions of the Yang-Mills equations (1) containing only the components of the SU(3)-field which are Coulomb-like or linear in r. Additionally we impose the Lorentz condition on the sought solutions. The latter condition is necessary for quantizing the gauge fields consistently within the framework of perturbation theory (see, e. g. Ref. [41]), so we should impose the given condition that can be written in the form $\mathrm{div}(A) = 0$, where the divergence of the Lie algebra valued 1-form $A = A_\mu dx^\mu = A^a_\mu \lambda_a dx^\mu$ is defined by the relation (see, e. g., Refs. [40])

$$\mathrm{div}\, A = *(d * A) = \frac{1}{\sqrt{\delta}}\partial_\mu(\sqrt{\delta} g^{\mu\nu} A_\nu)\,. \tag{D.2}$$

It should be emphasized that, from the physical point of view, the Lorentz condition reflects the fact of transversality for gluons that arise as quanta of SU(3)-Yang-Mills field when quantizing the latter (see, e. g., Ref. [41]).

We shall use the Hodge star operator action on the basis differential 2-forms on Minkowski spacetime with local coordinates t, r, ϑ, φ in the form

$$*(dt \wedge dr) = -r^2 \sin\vartheta d\vartheta \wedge d\varphi\,, *(dt \wedge d\vartheta) = \sin\vartheta dr \wedge d\varphi\,,$$

$$*(dt \wedge d\varphi) = -\frac{1}{\sin\vartheta} dr \wedge d\vartheta\,, *(dr \wedge d\vartheta) = \sin\vartheta dt \wedge d\varphi\,,$$

$$*(dr \wedge d\varphi) = -\frac{1}{\sin\vartheta} dt \wedge d\vartheta \, , *(d\vartheta \wedge d\varphi) = \frac{1}{r^2 \sin\vartheta} dt \wedge dr \, , \tag{D.3}$$

so that on 2-forms $*^2 = -1$. More details about the Hodge star operator can be found in [40].

The most general ansatz for a spherically symmetric solution is $A = A_t(r)dt + A_r(r)dr + A_\vartheta(r)d\vartheta + A_\varphi(r)d\varphi$. But then the Lorentz condition (D.2) for the given ansatz gives rise to

$$\sin\vartheta \partial_r(r^2 A_r) + \partial_\vartheta(\sin\vartheta A_\vartheta) = 0,$$

which yields $A_r = \frac{C}{r^2} - \frac{\cot\vartheta}{r^2} \int A_\vartheta(r) dr$ with a constant matrix C. But the confining solutions should be spherically symmetric and contain only the components which are Coulomb-like or linear in r, so one should put $C = A_\vartheta(r) = 0$. Consequently, the ansatz $A = A_t(r)dt + A_\varphi(r)d\varphi$ is the most general spherically symmetric one.

For the latter ansatz we have $F = dA + gA \wedge A = -\partial_r A_t dt \wedge dr + \partial_r A_\varphi dr \wedge d\varphi + g[A_t, A_\varphi] dt \wedge d\varphi$, where $[\cdot, \cdot]$ signifies matrix commutator.

Then, according to (D.3), we obtain

$$*F = (r^2 \sin\vartheta)\partial_r A_t d\vartheta \wedge d\varphi - \frac{1}{\sin\vartheta}\partial_r A_\varphi dt \wedge d\vartheta -$$
$$\frac{g}{\sin\vartheta}[A_t, A_\varphi] dr \wedge d\vartheta \, , \tag{D.4}$$

which entails

$$d * F = \sin\vartheta \partial_r(r^2 \partial_r A_t) \, dr \wedge d\vartheta \wedge d\varphi + \frac{1}{\sin\vartheta}\partial_r^2 A_\varphi \, dt \wedge dr \wedge d\vartheta \, , \tag{D.5}$$

while

$$*F \wedge A - A \wedge *F =$$
$$\left(r^2 \sin\vartheta[\partial_r A_t, A_t] - \frac{1}{\sin\vartheta}[\partial_r A_\varphi, A_\varphi] \right) dt \wedge d\vartheta \wedge d\varphi$$
$$-\frac{g}{\sin\vartheta}\Big([[A_t, A_\varphi], A_t] \, dt \wedge dr \wedge d\vartheta$$
$$+[[A_t, A_\varphi], A_\varphi] \, dr \wedge d\vartheta \wedge d\varphi \Big) . \tag{D.6}$$

Under the circumstances the Yang-Mills equations (D.1) are tantamount to the conditions

$$\partial_r(r^2\partial_r A_t) = -\frac{g^2}{\sin^2\vartheta}[[A_t, A_\varphi], A_\varphi], \tag{D.7}$$

$$\partial_r^2 A_\varphi = -g^2[[A_t, A_\varphi], A_t], \tag{D.8}$$

$$r^2\sin\vartheta[\partial_r A_t, A_t] - \frac{1}{\sin\vartheta}[\partial_r A_\varphi, A_\varphi] = 0. \tag{D.9}$$

The key equation is (D.7) because the matrices A_t, A_φ depend on merely r and (D.7) can be satisfied only if the matrices $A_t = A_t^a\lambda_a$ and $A_\varphi = A_\varphi^a\lambda_a$ belong to the so-called Cartan subalgebra of the SU(3)-Lie algebra. Let us remind that, by definition, a Cartan subalgebra is a maximal abelian subalgebra in the corresponding Lie algebra, i. e., the commutator for any two matrices of the Cartan subalgebra is equal to zero (see, e.g., Ref. [45]). For SU(3)-Lie algebra the conforming Cartan subalgebra is generated by the Gell-Mann matrices λ_3, λ_8 which are

$$\lambda_3 = \begin{pmatrix} 1 & 0 & 0 \\ 0 & -1 & 0 \\ 0 & 0 & 0 \end{pmatrix}, \lambda_8 = \frac{1}{\sqrt{3}}\begin{pmatrix} 1 & 0 & 0 \\ 0 & 1 & 0 \\ 0 & 0 & -2 \end{pmatrix}. \tag{D.10}$$

Under the situation we should have $A_t = A_t^3\lambda_3 + A_t^8\lambda_8$ and $A_\varphi = A_\varphi^3\lambda_3 + A_\varphi^8\lambda_8$, then $[A_t, A_\varphi] = 0$ and we obtain

$$\partial_r(r^2\partial_r A_t) = 0, \partial_r^2 A_\varphi = 0, \tag{D.11}$$

while (D.9) is identically satisfied and (D.11) gives rise to the solution (3) with real constants a_j, A_j, b_j, B_j parametrizing the solution which proves the uniqueness theorem of Section 2 for the SU(3) Yang-Mills equations.

The more explicit form of (3) is

$$\begin{aligned} A_t^3 &= [(a_2 - a_1)/r + A_1 - A_2]/2, \\ A_t^8 &= [A_1 + A_2 - (a_1 + a_2)/r]\sqrt{3}/2\,, \\ A_\varphi^3 &= [(b_1 - b_2)r + B_1 - B_2]/2, \\ A_\varphi^8 &= [(b_1 + b_2)r + B_1 + B_2]\sqrt{3}/2\,. \end{aligned} \tag{D.12}$$

Clearly, the obtained results may be extended over all SU(N)-groups with $N \geq 2$ and even over all semisimple compact Lie groups since for them the corresponding Lie algebras possess just the only Cartan subalgebra. Also we can talk about the compact non-semisimple groups, for example, U(N). In the latter case additionally to Cartan subalgebra we have centrum consisting from the matrices of the form αI_N (I_N is the unit matrix $N \times N$) with arbitrary constant α.

The most relevant physical cases are of course U(1)- and SU(3)-ones (QED and QCD). In particular, the U(1)-case allows us to build the classical model of confinement (see Section 2 and Ref. [46]).

At last, it should also be noted that the nontrivial confining solutions obtained exist at any gauge coupling constant g, i. e. they are essentially *nonperturbative* ones.

References

[1] Goncharov, Yu. P. *Mod. Phys. Lett.* 2001, A16, 557–569.

[2] Goncharov, Yu. P. *Phys. Lett.* 2005, B617, 67–77.

[3] Goncharov, Yu. P. In *New Developments in Black Hole Research*, Kreitler, P. V.; Ed.; Nova Science Publishers, New York, 2006; 67–121.

[4] Goncharov, Yu. P. *Europhys. Lett.* 2003, 62, 684–690.

[5] Goncharov, Yu. P.; Choban, E. A. *Mod. Phys. Lett.* 2003, A18, 1661–1671.

[6] Goncharov, Yu. P.; Bytsenko, A. A. *Phys. Lett.* 2004, B602, 86–96.

[7] Goncharov, Yu. P. *Nucl. Phys.* 2008, A808, 73–94.

[8] Goncharov, Yu. P. *Phys. Lett.* 2006, B641, 237–245.

[9] Goncharov, Yu. P. *Phys. Lett.* 2007, B652, 310–318.

[10] Goncharov, Yu. P. *Mod. Phys. Lett.* 2007, A22, 2273–2285.

[11] Goncharov, Yu. P. *J. Phys. G: Nucl. Part. Phys.* 2008, 35, 095006.

[12] Goncharov, Yu. P. *Nucl. Phys.* 2008, A812, 99–124.

[13] Goncharov, Yu. P. *Eur. Phys. J.* 2010, A46, 139–155.

[14] Wilson, K. *Phys. Rev.* 1974, D10, 2445–2459.

[15] Bander, M. *Phys. Rep.* 1981, 75, 205–286.

[16] Perkins, D. H. *Introduction to High Energy Physics*; Cambridge University Press, Cambridge, 2000.

[17] Nakamura, K. et al. (Particle Data Group) *J. Phys. G: Nucl. Part. Phys.* 2010, 37, 075021.

[18] Deur, A. *Nucl. Phys.* 2005, A755, 353.

[19] Deur, A. et al. *Phys. Lett.* 2007, B650, 244.

[20] Prudnikov, A. P.; Brychkov, Yu. A.; Marichev, O. I. *Integrals and Series. Elementary Functions*; Nauka, Moscow, 1981.

[21] Berestezkiy, V. B.; Lifshits, E. M.; Pitaevskiy, L. P. *Quantum Electrodynamics*; Fizmatlit, Moscow, 2002.

[22] Aliev, T. M. ; Kanik, I.; Savci, I. *Phys. Rev.* 2003, D68, 056002.

[23] Samsonov, A. *Yad. Fiz.* 2005, 68, 114.

[24] Burkardt, M.; Schnell, G. *Phys. Rev.* 2006, D74, 013002.

[25] Landau, L. D.; Lifshits, E. M. *Field Theory*; Nauka, Moscow, 1988; 3–460.

[26] Levich, V. G.; Vdovin, Y. A.; Myamlin, V. A. *Course of Theoretical Physics*; Nauka, Moscow, 1971; Vol. 2, 3–936.

[27] Landau, L. D.; Lifshits, E. M. *Quantum Mechanics. Nonrelativistic Theory*; Nauka, Moscow, 1989; 3–768.

[28] Dahiya, H.; Gupta, M. *Phys. Rev.* 2003, D67, 114015.

[29] Hey, A. J. G.; Kelly, R. L. *Phys. Rep.* 1983, 96, 71.

[30] Klempt, E.; Richard, J-M. *Rev. Mod. Phys.* 2010, 82, 1095.

[31] Farhi, E.; Susskind, L. *Phys. Rep.* 1981, 74, 277.

[32] Kaul, R. K. *Rev. Mod. Phys.* 1983, 55, 449.

[33] Cheng, T.-P.; Li, L.-F. *Gauge Theory of Elementary Particle Physics*; Clarendon Press, Oxford, 1984.

[34] Hirn, J.; Martin, A.; Sanz, V. *Phys. Rev.* 2008, D78, 075026.

[35] Dugne, J. -J.; Fredriksson, S.; Hansson, J. *Europhys. Lett.* 2002, 57, 188.

[36] Jaffe, A.; Witten, E. In *The Millennium Prize Problems*; Carlson, J.; Jaffe, A.; Wiles, A.; Eds.; American Mathematical Society: Providence, RJ, 2006; 129–152.

[37] Goncharov, Yu. P. *Pis'ma Zh. Exp. Teor. Fiz.* 1999, 69, 619–625.

[38] Goncharov, Yu. P. *Phys. Lett.* 1999, B458, 29–35.

[39] Vilenkin, N. Ya. *Special Functions and Theory of Group Representations*; Nauka, Moscow, 1991.

[40] *Géométrie Riemannian en Dimension 4. Seminaire Arthur Besse*; Cedic/Fernand Nathan, Paris, 1981.

[41] Ryder, L. H. *Quantum Field Theory*; Cambridge University Press, Cambridge, 1985.

[42] Besse, A. L. *Einstein Manifolds*; Springer-Verlag, Berlin, 1987.

[43] Lawson, H. B., Jr.; Michelsohn, M.-L. *Spin Geometry*; Princeton University Press, Princeton, 1989.

[44] Postnikov, M. M. *Riemannian Geometry*; Factorial, Moscow, 1998.

[45] Barut, A. O.; Raczka, R. *Theory of Group Representations and Applications*; Polish Science, Warszawa, 1977.

[46] Goncharov, Yu. P.; Firsova, N. E. *Int. J. Theor. Phys.* 2010, 49, 1155–1173.

[47] Goncharov, Yu. P. *Int. J. Theor. Phys.* 2012, 51, 428–437.

In: Recent Advances in Quarks Research
Editors: H. Fujikage and K. Hyobanshi
ISBN: 978-1-62257-970-9
© 2013 Nova Science Publishers, Inc.

Chapter 4

THERMODYNAMICAL DESCRIPTION OF HADRON-QUARK PHASE TRANSITION AND ITS IMPLICATIONS ON COMPACT-STAR PHENOMENA

Nobutoshi Yasutake[1,*], Tsuneo Noda[2], Hajime Sotani[3], Toshiki Maruyama[4] and Toshitaka Tatsumi[5]

[1]Department of Physics, Chiba Institute of Technology, 2-1-1 Shibazono, Narashino, Chiba, Japan

[2]Department of Physics, Kyushu University, 6-10-1 Hakozaki, Higashi-ku, Fukuoka, Japan

[3]Division of Theoretical Astronomy, National Astronomical Observatory of Japan, 2-21-1 Osawa, Mitaka, Tokyo, Japan

[4]Advanced Science Research Center, Japan Atomic Energy Agency, Shirakata Shirane 2-4, Tokai, Ibaraki, Japan

[5]Department of Physics, Kyoto University, Kyoto, Japan

Abstract

One of the most promising possibilities may be the appearance of quark matter in astrophysical phenomena in the light of recent progress in observations. The mechanism of deconfinement is not well understood, but the thermodynamical aspects of the hadron-quark (HQ) phase transition have

*E-mail address: nobutoshi.yasutake@it-chiba.ac.jp

been extensively studied in recent years. Then the mixed phase of hadron and quark matter becomes important; the proper treatment is needed to describe the HQ phase transition and derive the equation of state (EOS) for the HQ matter, based on the Gibbs conditions for phase equilibrium. We here adopt a EOS based on the baryon-baryon interactions including hyperons for the hadron phase, while we use rather simple EOS within the MIT bag model for the quark phase. One of the interesting consequences may be the appearance of the inhomogeneous structures called "pasta", which are brought about by the surface and the Coulomb interaction effects. We present here a comprehensive review of our recent works about the HQ phase transition in various astrophysical situations: cold catalyzed matter, hot matter and neutrino-trapped matter. We show how the pasta structure becomes unstable by the charge screening of the Coulomb interaction, thermal effect or the neutrino trapping effect. Such inhomogeneous structure may affect astrophysical phenomena through its elasticity or thermal properties. Here we also discuss some implications on supernova explosion, gravitational wave and cooling of compact stars.

1. Introduction

The mechanism of the hadron-quark (HQ) deconfinement is one of hot topics in modern physics. Theoretically, numerical simulations based on the lattice gauge theory have been done to estimate the critical temperature and figure out properties of the phase transition[1]. However, at present, these attempts do not provide definite results for the deconfinement transition in high density region. On the other hand, there have been done many theoretical studies by using the effective models of QCD[2, 3]. Therefore, one of the most hopeful strategies to get insight into this mechanism is to find or predict some signals about QCD phase transition in the astrophysical phenomena such as supernovae, black hole formations, mergers of neutron stars, cooling of compact stars, etc., in the light of recent theoretical and observational progress. Then the equation of state (EOS) is a basic element to discuss the implications on above astrophysical phenomena. There have been proposed and discussed various types of scenario concerned with the HQ deconfinement transitions at high-density region, but it is still unclear, even, whether the phase transition is the cross over or the first order [4]. Only assumption in this review is that the phase transition is of the first order as suggested by many model studies [2]. One of the direct consequences of this assumption is the emergence of the hadron-quark (HQ) mixed phase during the phase transition. In this review we first discuss the thermodynamical aspects of the HQ transition, and present some implications on astrophysical phenomena subsequently.

Historically, the mixed phases have been often treated very naively by applying the Maxwell construction to get the EOS in thermodynamic equilibrium. In 1990 Glendenning has pointed out that the phase equilibrium in multi-component system or in the case of plural chemical potential corresponding to the conserved quantities such as in neutron star matter must be considered based on the more general Gibbs conditions [5, 6]. He emphasized that the neutron-star matter should be treated as a *binary* system specified by two conserved quantities, electromagnetic charge and baryon number. Thus the pressure becomes no more constant as in the Maxwell construction, but still increases in the mixed phase. He also demonstrated that the mixed phase spreads in a rather wide density region within a bulk calculation, where two kinds of semi-infinite matter are considered in phase equilibrium, and the pressure balance, chemical equilibrium and the global charge-neutrality conditions are imposed, discarding the Coulomb interaction. It is important to see that particle fractions in each phase are no more the same for given average particle-fractions. Generally any phase transition in a system with two or more conserved charges must be *non-congruent* [1] . However, the Coulomb interaction is discarded in this calculation and we can by no means estimate how important it is, or the surface energy can not be taken into account a priori. If we consider a more realistic situation, we must consider inhomogeneous matter with various geometric structures called "*pasta*", taking into account the finite-size effects such as the Coulomb interaction and the surface tension. The pasta resembles the *nuclear pasta* in liquid-gas transition at subnuclear densities [7, 8, 9, 10, 11, 12]. Heiselberg et al. have demonstrated that such finite-size effects are important for the mixed phase in the context of the HQ transition [13]. Treating the pasta structure as in ref. [14], they simply considered uniform particle densities in both phases. They showed a possibility of that the finite-size effects may largely restrict the region of the mixed phase. Subsequently, Voskresensky et al. have shown that the rearrangement of particle densities and the screening effect for the Coulomb interaction are important, and the large surface tension gives the mechanical instability of the geometrical structures in the pasta phase [15]. Using the effective model for hadron matter and the MIT bag model for quark matter, Endo et al. explicitly demonstrated these features [16]. Following this idea, we have studied the

[1]When one dare to apply the Maxwell construction in such systems, one must violate the Gibbs conditions by ,e.g., imposing local charge neutrality.

pasta structures brought by the kaon condensation and the HQ deconfinement transition [17, 16, 18]. More recently we have extended our framework to deal with the finite temperature case or the neutrino-trapping matter, which may be relevant to supernova explosions or neutron star mergers [19, 20].

In this review we first present thermodynamic concepts of the non-congruent transition of the HQ phase transition (Section 2). In the section we will also present some results about the phase transition, where the effects of finite temperature or neutrino trapping are also discussed. In Section 3 is discussed some astrophysical implications of the HQ phase transition. Section 4 is devoted to summary and concluding remarks.

2. Phase Equilibrium in Thermodynamics on Hadron-Quark Phase Transition

In this section we discuss the emergence of the pasta structure associated with the HQ deconfinement transition[2]. For our purpose we need both EOS's of hadron and quark matter as realistically as possible. As we mentioned in the last section, no one knows how to exactly calculate the HQ phase transition at high density region. The studies by using the effective models of QCD such as the MIT bag model or the NJL model have been actively done, instead. We, hereafter, use the MIT bag model for simplicity.

We adopt a realistic EOS for the hadron matter based on the microscopic nucleon-nucleon (NN), and nucleon-hyperon (NY) interactions. Although there are many models to describe the EOS with hyperons, they have many uncertainties which comes from the uncertainty of NY and YY interactions. However, nowadays, the potentials of the baryon-baryon interactions have been directly calculated by the lattice QCD calculations [22]. We hope that they provide realistic NY and YY interactions in near future. Motivated by such background, it becomes a very hot topic to calculate the EOS from the baryon-baryon interaction directly. Here, we show the EOS by the theoretical framework of the nonrelativistic Brueckner-Hartree-Fock (BHF) approach [23] based on the microscopic NN and NY interactions. The BHF calculation is a reliable and well-controlled theoretical approach for the study of dense baryon matter.

In Fig. 1, we sum up our strategy to construct EOS including the HQ phase for the astrophysical implications. Since our goal is to figure out the realistic

[2]We, hereafter, dismiss the possibility of color superconductivity [21].

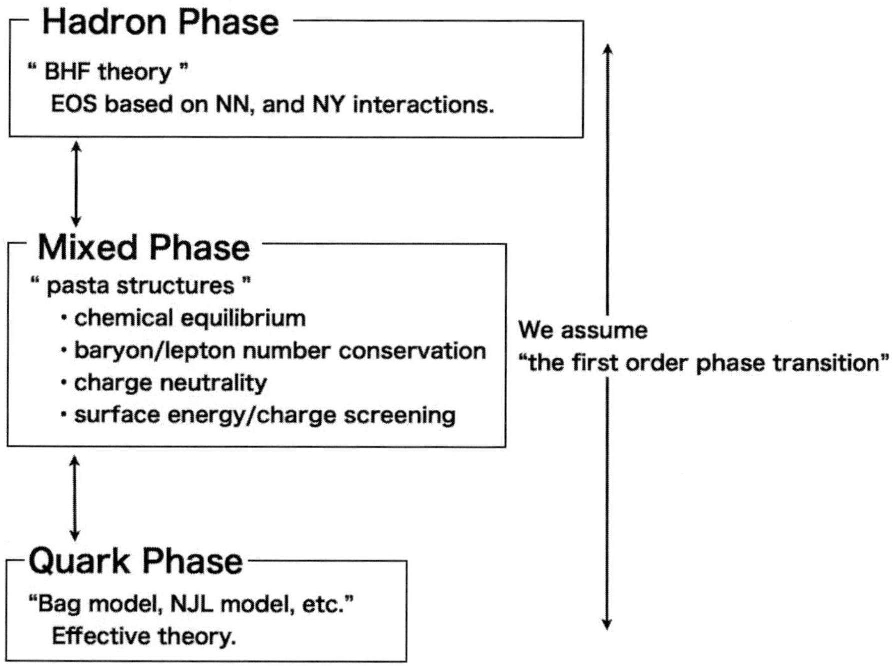

Figure 1. Schematic picture of our strategy on the study about the HQ phase transition for the astrophysical implications.

properties of the HQ phase transition, we chose one of the most realistic ways to know the HQ mixed phase at high-density region; NY and YY interactions will be clear in the future for the hadron phase.

Then we consider the phase-equilibrium by including the finite-size effects. The Coulomb interaction and the surface energy are taking into account in a consistent way: charge screening effect is automatically took into account by solving the Poisson equation, but the surface energy at the interface of hadron and quark matter is treated in a phenomenological way by using the surface tension parameter. Thus, the largest ambiguity in our study is the uncertainty in the quark models. Hence, one of the main purposes of our study is to avoid this ambiguity as much as possible by comparing our results with astrophysical phenomena. Here we sum up our method, and report our current status.

2.1. Hadron Phase; The Brueckner-Hartree-Fock Theory

BHF theory is well-known to describe hadron matter using the realistic baryon-baryon interaction. In this review we do not explain the details of the procedure, but they can be found in Refs. [23, 24, 25, 26]. We adopt the Argonne V_{18} potential [27] for NN interaction, and semi-phenomenological Urbana UIX nucleonic three-body forces [29] and the Nijmegen soft-core NSC89 NY potentials [30]. Unfortunately, there are not reliable potentials for the hyperon-hyperon interaction now. Therefore, we neglect them here.

To take into account the effects of finite temperature, we adopt *Frozen Correlations Approximation* [31, 32, 33]. The correlations at finite temperature are assumed to be the same with the ones at zero temperature. It is found to be a good approximation at finite temperature as Refs. [31, 32, 33].

In the following, we briefly describe our framework. First, we get the chemical potential μ_i from the number density n_i,

$$n_i = \frac{g}{(2\pi)^3}\int_0^\infty f_i(p)\,4\pi p^2 dp \tag{1}$$

$$f_i(p) = \frac{1}{\exp\{(\varepsilon_i-\mu_i)/T\}+1}, \tag{2}$$

where ε_i and $f_i(p)$ are the single-particle energy and the Fermi-Dirac distribution function, respectively, whereas subscript i denotes the particle species, $i = n, p, \Lambda, \Sigma^-$. We set each degeneracy factor $g = 2$, and adopt each mass as $m_n = m_p = 939$ MeV, $m_\Lambda = 1115.7$ MeV, and $m_{\Sigma^-} = 1197.4$ MeV. Note that ε_i includes the interaction energy U_i as well as the kinetic energy [31, 34],

$$\varepsilon_i = \sqrt{m_i^2+p^2}+U_i. \tag{3}$$

Finally, we get the free-energy density $\mathcal{F}$ as follows,

$$\mathcal{F}_H = \sum_i\left[\frac{g}{(2\pi)^3}\int_0^\infty\sqrt{m_i^2+p^2}f_i(p)\,4\pi p^2dp+\frac{1}{2}U_in_i\right]-Ts_H, \tag{4}$$

where s_H is the entropy density calculated from

$$s_H = -\sum_i\frac{g}{(2\pi)^3}\int_0^\infty\{f_i(p)\ln f_i(p)+(1-f_i(p))\ln(1-f_i(p))\}\,4\pi p^2dp. \tag{5}$$

The total pressure for the uniform hadron phase is given by

$$P_H = \sum_i \mu_i n_i - \mathcal{F}_H. \tag{6}$$

2.2. Quark Phase; The Bag Model

Next, we explain how to treat the quark phase based on the thermodynamic bag model. In this model, we get the number density n_Q, the pressure P_Q, and the energy density ϵ_Q,

$$n_Q = \sum_q \frac{g}{(2\pi)^3} \int_0^\infty f_q(p) 4\pi p^2 dp, \tag{7}$$

$$\epsilon_Q = \sum_q \frac{g}{(2\pi)^3} \int_0^\infty \sqrt{m_q^2 + p^2} f_q(p) 4\pi p^2 dp + B, \tag{8}$$

$$s_Q = -\sum_q \frac{g}{(2\pi)^3} \int_0^\infty \{f_q(p)\ln f_q(p) + (1 - f_q(p))\ln(1 - f_q(p))\} 4\pi p^2 dp \tag{9}$$

$$\mathcal{F}_Q = \epsilon_Q - T s_Q \tag{10}$$

$$P_Q = \sum_q \mu_q n_q - \mathcal{F}_Q. \tag{11}$$

for uniform quark matter, where $f_q(p)$ is the Fermi-Dirac distribution function of the quark q $(= u, d, s)$, m_q its current mass, and B the energy-density difference between the perturbative vacuum and the true vacuum, i.e., the bag constant is set to be 100 MeV·fm^{-3}. The baryon density n_B is given by $(n_u + n_d + n_s)/3$.

Since there are still many uncertainties for the vacuum property, we here also adopt another type of B by the density dependent bag model [33]. In this model, we set the vacuum energy $B(n_B)$ as

$$B(n_B) = B_\infty + (B_0 - B_\infty) \exp\left[-\beta \frac{n_B}{n_0}\right]$$

with $B_\infty = 50$ MeV·fm^{-3}, $B_0 = 400$ MeV·fm^{-3}, and $\beta = 0.17$. Here, we set the saturation density as $n_0 = 0.17$ fm^{-3}. This model has been proposed in Ref. [33], on the basis of experimental results obtained at CERN SPS on formation of a quark-gluon plasma.

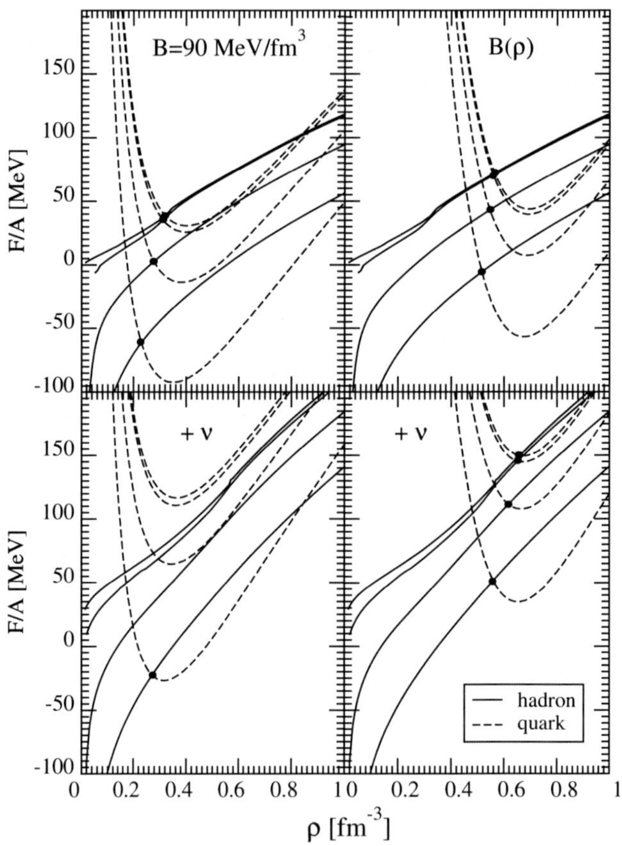

Figure 2. Figure from Ref.[33]. Free energy per baryon as a function of baryon density of beta-stable hadron matter (solid lines) and quark matter (dashed lines) with a bag constant $B = 90$ MeV/fm^3 (left panels) or a density-dependent bag parameter (right panels) with (lower panels) and without (upper panels) neutrino trapping at different temperatures $T = 0$, 10, 30, 50 MeV (upper to lower curves).

Here, we show free energy per baryon F/A as a function of baryon density for hadron matter and quark matter at different temperature ($T = 0, 10, 30, 50$ MeV) in Fig. 2. We adopted the thermodynamic bag model ($B = 90$ MeV/fm^3) as the quark matter for left panels, but the density dependent bag model for the right panels. The lower panels show the case with neutrino trapping, $Y_L = 0.4$, while upper panels for without. Note that the values of F/A of hadrons are lower than the ones of quark matter at neutrino trapped case at $T = 0, 10, 30$ for any density. Therefore it is not appropriate for our assumption that the HQ transition of the first order occurs at high density region. Hence we adopt the density-dependent bag model for the lepton trapping case in this article.

We note that the bag model should be quite simple to describe the quark matter. We will update the quark model to more sophisticated models such as NJL model [35, 36, 37], pNJL model [38, 39, 40, 41, 42, 43], or Dyson-Schwinger model [44], etc..

2.3. The Maxwell Construction and the Gibbs Conditions

2.3.1. Hadron-Quark Phase Transition as a Pure System Under the Maxwell Construction

In this subsection, we introduce the mixed phase to discuss the phase equilibrium between hadron and quark phases. We can expect the HQ transition to appear in various circumstance, supernovae, protoneutron stars, cold neutron stars or neutron-star mergers, where temperature may become high or neutrinos may be trapped. So we consider here a general situation, where exist hadrons such as nucleons and hyperons and leptons such as electrons and neutrinos.

First, we consider the HQ phase transition without considering the finite-size effects which is called the bulk calculation. The naive treatment of the phase equilibrium in the first-order phase transition should be the application of the Maxwell construction. In this treatment for neutron-star matter we implicitly assume that electric charge is always locally neutralized in each phase and lepton number is equal to each other; only baryon number is different between two phases. So the Maxwell construction means that the matter is regarded as a pure system, discarding the electromagnetic charge and lepton number.

Relaxing the local charge-neutrality condition and equality of lepton number, we must impose the Gibbs conditions in the proper treatment of the HQ transition in neutron stars. In this treatment, we impose the balance of the chemical potentials of quarks and baryons as well as the pressure balance and thermodynamical equilibrium between two phases. The latter two conditions are obvious,

$$P_H = P_Q, \qquad T_H = T_Q, \tag{12}$$

where P_H (P_Q), T_H (T_Q) are pressure and temperature of hadron (quark) phase. The former conditions are reduced to only one within the Maxwell construction for the baryon-number chemical potential μ_B [3],

$$\mu_B(\text{hadron phase}) = \mu_B(\text{quark phase}), \tag{13}$$

where the baryon chemical potential in the hadron (quark) phase is given by $\mu_B = \mu_n$ ($\mu_B = \mu_u + 2\mu_d$). Here, μ_n, μ_u, and μ_d is neutron, u-quark, and d-quark chemical potential, respectively.

We also assume the beta equilibrium among the particles as

$$\mu_u = \frac{1}{3}\mu_B + \frac{2}{3}\mu_C^Q, \qquad \mu_d = \mu_s = \frac{1}{3}\mu_B - \frac{1}{3}\mu_C^Q, \tag{14}$$

for the quark phase,

$$\mu_n = \mu_\Lambda = \mu_B, \qquad \mu_p = \mu_B + \mu_C^H, \qquad \mu_{\Sigma^-} + \mu_p = 2\mu_B, \tag{15}$$

for the hadron phase, where the quark chemical potentials are shown as $\mu_{u,d,s}$ for u, d, s-quarks, the hadron chemical potentials $\mu_{n,p,\Lambda,\Sigma^-}$ for n, p, Λ, Σ^-. In these equations, we have also introduced the two kinds of variables to understand the mechanism of the HQ phase transition systematically; the lepton chemical potential $\mu_L^{H(Q)}$ and the charge chemical potential $\mu_C^{H(Q)}$ are given as

$$\mu_L^{H(Q)} = \mu_{\nu_e}^{H(Q)}, \qquad \mu_C^{H(Q)} = \mu_L^{H(Q)} - \mu_e^{H(Q)}, \tag{16}$$

in hadron (quark) matter. Here the lepton chemical potentials are shown as $\mu_e^{H(Q)}$ and $\mu_{\nu_e}^{H(Q)}$ for electrons and electron type neutrinos, respectively. Note that μ_L and μ_C are not equal to each other between two phases; $\mu_L^H \neq \mu_L^Q$ and $\mu_C^H \neq \mu_C^Q$. Under this construction, we assume only the equality of baryon chemical potentials μ_B between two phases shown as Eq.(13). Hence the matter is regarded as a pure system in this treatment. As a result, the EOS has a density jump in general.

[3]Another version of the Maxwell construction may be possible [45], where the effects of the lepton chemical potentials on μ_B are taken into account.

2.3.2. Hadron-Quark Phase Transition as a Ternary System Under the Gibbs Conditions

So far the charge equilibrium is completely discarded between two phases: the above assumption uses the charge neutrality condition in the local form, while it must be attained as a whole system. Consequently, the Maxwell construction arises as a consequence of further strong constraint. In such treatment, the lepton number is also assumed to be conserved locally. In the proper treatment of the HQ transition we must consider both phases as ternary systems specified by the three conserved quantities, electromagnetic charge, lepton number, as well as baryon number.

In other words, we must add two chemical equilibrium-conditions into Eq.(13) as

$$\mu_B^H = \mu_B^Q, \qquad \mu_L^H = \mu_L^Q, \qquad \mu_C^H = \mu_C^Q. \tag{17}$$

Here, there are three identical components; the baryon number (which corresponds to μ_B), charge (μ_C), and lepton number (μ_L). The Maxwell construction is then meaningless and we must use the Gibbs conditions to describe the phase equilibrium in a thermodynamically consistent way. Hence, the EOS dose not have density jump although we assume that the phase transition is of the first order.

Since we impose the global charge neutrality (GCN), the total charge as well as baryon-number are shared by the two phases, conservation should be satisfied as following;

$$n_B = \chi n_B^Q + (1-\chi) n_B^H \ , \tag{18}$$

$$Y_C\, n_B = \chi\, Y_C^Q\, n_B^Q + (1-\chi)\, Y_C^H\, n_B^H \ , \tag{19}$$

$$Y_L\, n_B = \chi\, Y_L^Q\, n_B^Q + (1-\chi)\, Y_L^H\, n_B^H \ , \tag{20}$$

where n_B, n_B^Q, n_B^H, and χ are respectively the total baryon number density, the baryon number density in hadron matter, the effective baryon number density in quark matter, and the volume fraction which indicates the volume of the quark phase divided by the total volume. Here, n_B^Q is one third of the total quark number density. The values of Y_C and Y_L are the total fraction of positive particles and leptons, whereas $Y_{C(L)}^Q$ and $Y_{C(L)}^H$ denote the fractions of positive particles (leptons) per baryon in quark matter, and in hadron matter. Note that

$Y^Q_{C(L)} = Y^H_{C(L)}$ are assumed in the previous treatment (13) instead of $\mu^Q_{C(L)} = \mu^H_{C(L)}$.

2.3.3. Hadron-Quark Phase Transition as a Binary System with the Local Charge Neutrality

Recently the hybrid types of the Maxwell construction and the Gibbs conditions have been suggested, depending on the physical situation [45]. One of them is the Gibbs conditions assuming the local charge neutrality (LCN), $Y^Q_C = Y^H_C = 0$. In this treatment, the chemical equilibriums are described as follow;

$$\mu^H_B = \mu^Q_B, \qquad \mu^H_L = \mu^Q_L. \tag{21}$$

We shall see LCN is turned out to be achieved in some cases as in the protoneutron stars (subsection 2.8) .

In the last of this subsection, we show some examples of EOS adopted the Gibbs conditions for the HQ mixed phase in Fig.3, where we compare both conditions of GCN and LCN. The left panel is at finite isentropic case $s = 2\ [/k_B]$, and $Y_L = 0.35$ assuming the matter in protoneutron stars. One can see that the proper treatment of the Gibbs conditions (GCN) gives almost the same result as that in the hybrid treatment of the Gibbs conditions with a further constraint, LCN, in this case. The right panel is at zero temperature without neutrinos assuming the case in the neutron stars. Note that, without neutrinos, the Gibbs conditions with LCN give completely the same result with the Maxwell construction. Consequently, the EOS has a density jump as shown in the right panel in Fig.3 while pressure smoothly increases in GCN. We also show the uniform EOSs of quark matter and hadron matter in these figures.

2.4. Finite-Size Effects and "Pasta" Structures

We have presented general features of the non-congruent transition in a clear fashion with a help of the bulk calculation, discarding the Coulomb interaction and surface effects called the finite-size effects. However, the bulk calculation is insufficient to figure out the essential aspects of the non-congruent transition.

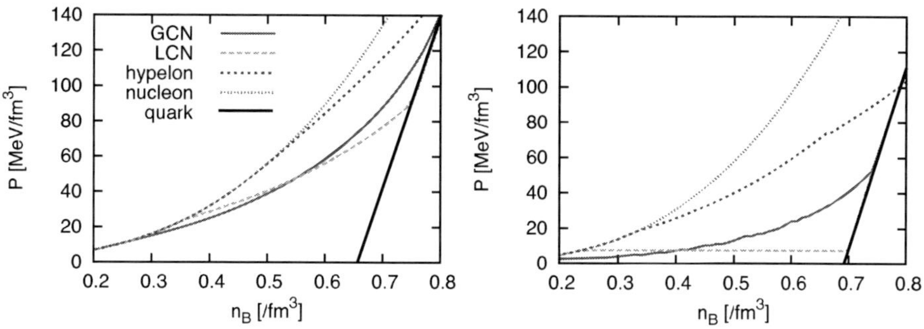

Figure 3. Equation of sate considering the mixed phases imposed the Gibbs conditions assuming GCN and LCN, which is labeled or 'GCN' and 'LCN'. For comparison, we also shows the uniform matter of quark matter (labeled as 'quark') and hadron matter. Here, the EOS labeled as 'hyperon' shows the hadron matter including hyperons, while the EOS labeled as 'nucleon' is without hyperons. Left panel shows the EOS at isentropic, $s = 2[/k_B]$, and at $Y_L = 0.35$. Right panel is the same with left one, but for at zero temperature without neutrinos.

Actually we can see the emergence of imhomogeneous structure with various geometrical shapes, called pasta (see Fig. 4), by the cooperative effect of the Coulomb interaction and the surface energy, which is similar to the nuclear pasta at subnuclear densities [7, 8, 9, 10, 11, 12]. The charge screening effect, which is a typical many-body effect in the presence of the Coulomb interaction, plays an important role in this context to make the pasta structures unstable, which is called mechanical instability [15]. The evaluation of the free energy of the pasta structures can be carried out by using the Wigner-Seitz approximation; dividing the whole space into equivalent unit cells, we solve the coupled equations for density profiles of particles and the Poisson equation in a selfconsistent way within the unit cell [46]. We further use the Thomas-Fermi (local density) approximation for each particle.

Assuming various geometrical structures, we must numerically search the minimum of the free energy

$$F = E_V + F_e + F_{\nu_e} + E_C + E_S \tag{22}$$

changing the volume ratio of quark matter to hadron matter, and the size of

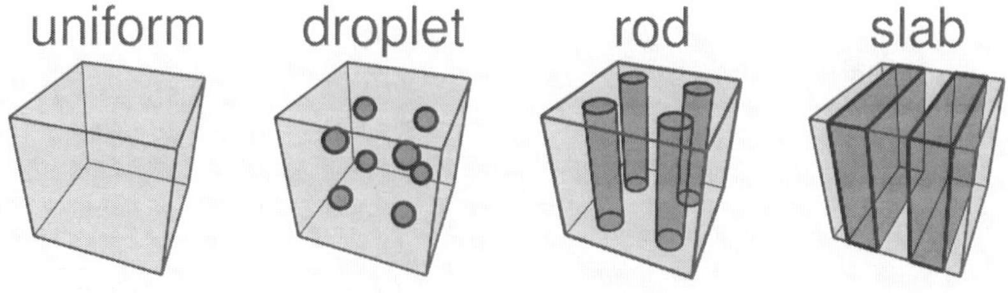

Figure 4. Schematic picture of "pasta" structures. When red phase appears in the blue phase, the red phase form droplets in the sea of blue. With increase of the volume fraction of red phase in the blue phase, the shape of the red phase changes from droplet to rod, slab, tube, and to bubble before pure red phase appears.

Wigner-Seitz cell R_W. Here the bulk energy E_V gives a main contribution,

$$E_V = \int_{V_H} dr^3 \mathcal{F}_H[n_i] + \int_{V_Q} dr^3 \mathcal{F}_Q[n_a], \tag{23}$$

where $\mathcal{F}_H$ and $\mathcal{F}_Q$ comes from the kinetic energy and strong interaction. The free energies of electrons, F_e, and neutrinos, F_{ν_e}, are simply calculated by the use of the Fermi-Dirac distribution functions. E_C denotes the interaction energy due to the Coulomb force. The treatment of the Coulomb potential may deserve some discussions, since it includes a delicate problem and sometimes has given rise to a confusion [15]. First of all we must notice that charge chemical potential μ_C itself is not physical quantity since it is gauge dependent in the presence of the Coulomb interaction $V_{\rm Coul}$. Only their combination of $\mu_C - V_{\rm Coul}$ is gauge invariant. Consider, for example, electrons without any neutrino. Then the electron chemical potential μ_e is equal to μ_C, so that the Gibbs condition means μ_e equals to each other in the mixed phase. On the other hand, the electron number density n_e should be represented as $(\mu_e - V_{\rm Coul})^3/(3\pi^2)$ in the gauge invariant way. Thus if $V_{\rm Coul}$ spatially varies in the mixed phase, each phase can have different number of electrons even if μ_e is the same in the both phases. In the extreme case $V_{\rm Coul}$ may resemble the step function and we have a simple picture of uniform electron density in both phases with different numbers. Note that there arises no such problem in the absence of the Coulomb potential as in the bulk calculation: the difference of the electron number directly reflects in the difference of the electron chemical potential in this case.

Here, we do not consider all kinds of anti-particles for simplicity. After the minimization of the free energy, we can get the thermodynamic quantities, such as pressure, entropy, etc.. The notation E_S stands for the surface energy which comes from a sharp interface between the hadron and quark phases with a fixed surface tension σ. Although the surface tension of the HQ interface is poorly known, some theoretical estimates based on the MIT bag model for strangelets [47] and lattice gauge simulations at finite temperature [48] have suggested a range of $\sigma \approx 10\text{–}100\ \text{MeV} \cdot \text{fm}^{-2}$. We will discuss the effects of the surface tension on the pasta structures in subsection 2.6.

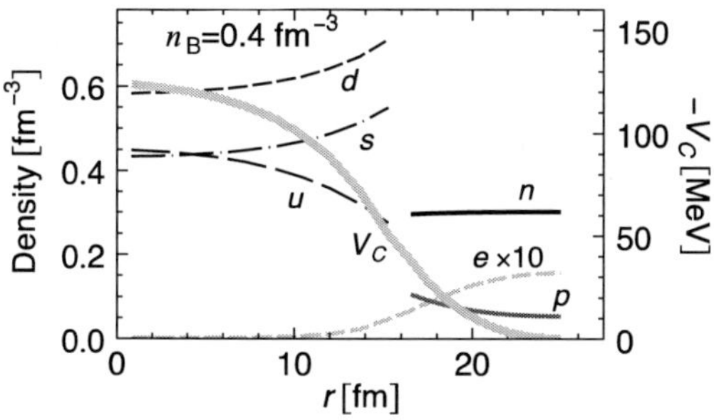

Figure 5. Density profiles and the Coulomb potential V_C within a 3D (quark droplet) Wigner-Seitz cell of the mixed phase at $n_B = 0.4\ \text{fm}^{-3}$. The cell radius and the droplet radius are $R_W = 26.7$ fm and $R = 17.3$ fm, respectively.

Fig. 5 shows an example of the resulting density profiles in the droplet phase for $n_B = 0.4\ \text{fm}^{-3}$ at zero temperature [18]. One can see the non-uniform density distributions of particle species together with the non-vanishing Coulomb potential. The quark phase is negatively charged, so that d and s quarks are repelled to the phase boundary, while u quarks gather at the center. The protons in the hadron phase are attracted by the negatively charged quark phase, while the electrons are repelled.

2.5. Hyperon Suppression in the Mixed Phase

Though we allow the freedom of hyperon mixture, hyperons do not actually come into the mixed phase. The suppression of hyperon mixture in the mixed phase is due to the fact that the hadron phase is positively charged. As shown in

Fig. 6, hyperons (Σ^-) appear in charge-neutral hadronic matter at low density to reduce the Fermi energies of electron and neutron. Without the charge-neutrality condition, on the other hand, there is symmetric nuclear matter at lower density and hyperons will be mixed there at higher density due to the large hyperon masses. Generally speaking hyperons are hardly mixed in the positively charged matter. Thus, the mixture of hyperons is suppressed in the mixed phase where the hadron phase is positively charged.

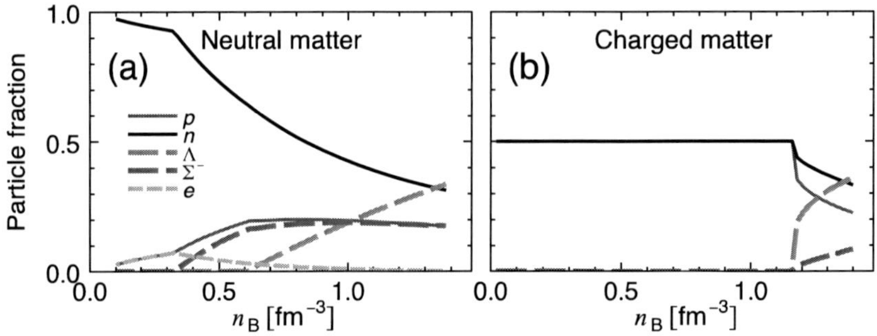

Figure 6. (a) Particle fractions of neutral matter with electrons. (b) The same quantity for charged matter without electrons, the low-density part of which corresponds to symmetric nuclear matter.

2.6. Effect of the Surface Tension

Let us consider the role of the surface tension on the mixed phase. As already mentioned, the surface tension in the interface of the hadron and quark phases is poorly known. If one uses a smaller surface tension parameter σ, the energy gets lower and the density range of the mixed phase gets wider. The limit of $\sigma = 0$ MeV$\cdot$ fm^{-2} leads to a bulk application of the Gibbs conditions without the Coulomb and surface effects. On the other hand, using a larger value of σ, we find that the scale of the inhomogeneity becomes large and the EOS gets closer to that of the Maxwell construction case. Beyond a limiting value of $\sigma \approx 65$ MeV $\cdot$ fm^{-2} the structure of the mixed phase becomes mechanically unstable [15]: for a fixed volume fraction $(R/R_W)^3$ the optimal values of R and R_W diverge and local charge neutrality is recovered in the mixed phase, where the energy density equals to that of the Maxwell construction (see Fig. 7).

Note that these roles of surface tension on the EOS is remarkable for the matter without neutrinos, i.e. a binary system, where the pasta structure is rather stable. If we consider the neutrino contamination, the matter should be treated as a ternary system. As we will discuss at subsection 2.8, the EOS could not be close to the one of the Maxwell construction, but to the one with the bulk Gibbs conditions (LCN) depending on the neutrino fraction.

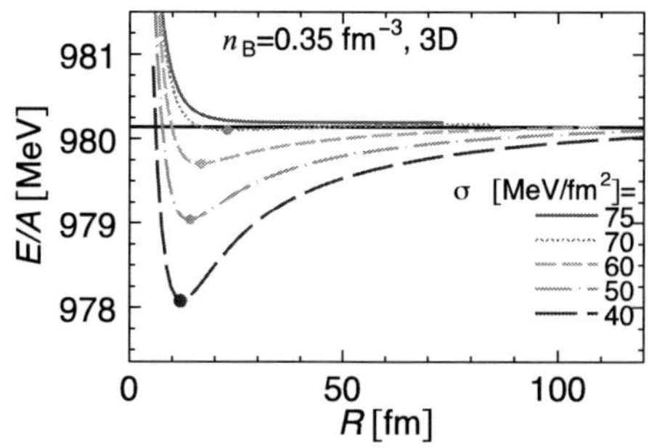

Figure 7. Droplet radius (R) dependence of the energy per baryon for fixed baryon density $n_B = 0.35\,\text{fm}^{-3}$ and different surface tensions. The temperature T is zero for all cases. The quark volume fraction $(R/R_W)^3$ is fixed for each curve. Dots on the curves show the local energy minima. The black line shows the energy of the Maxwell construction case.

2.7. Thermal Effects

In following we discuss the thermal and neutrino trapping effects on the HQ phase transition. In the left panel of Fig. 8 we show the dependence of the isothermal free-energy per baryon on the cell size in Eq. (22) for several temperatures at $n_B = 2\,n_0$, using the constant bag constant and the surface tension as $B = 100$ MeV$\cdot$fm^{-3} and $\sigma = 40$ MeV$\cdot$fm^{-2}, respectively [19]. Note that the free energy per baryon $\Delta F/A$ is normalized value by subtracting the one at infinite R_W, for the comparison of thermal effects. Here we assume the geometrical structure is the droplet and do not take into account neutrinos to see the genuine thermal effects. The quark volume fraction $(R/R_W)^3$ is fixed to be the optimal value at $T = 0$ MeV for each curve. From the this panel, we can see that the minimum point is shifted to a larger value of R as temperature is increased, and eventually disappears for $T > 60$ MeV. The temperature dependence of the free energy comes from the Coulomb energy, the surface energy and the correlation energy. The correlation energy is given by the difference in the bulk energy E_V in Eq. (22) [15], defined by

$$E_{\rm corr} \equiv E_V(n_i(\mathbf{r})) - E_V(n_i^{I,II}). \tag{24}$$

The rearrangement of the particle densities also induces a change in the volume energy E_V, which is composed of kinetic energy and strong-interaction energy. In the above calculation E_V itself is intrinsically large but independent on R, since we have assumed uniform densities $n_i^{I,II}$ in each phase. Once the Coulomb interaction is properly taken into account, particle densities are no more uniform to produce a R dependence in E_V. The detailed analysis showed that $E_{\rm corr}$ gives negative values and behaves R^{-1} for large R. By comparing the temperature dependence of these contributions, we can see that the correlation energy is primarily responsible to the behavior of the minimum point. It means that the mixed phase becomes less stable as temperature is increased [19].

In the right panel of Fig. 8, we show the EOS (pressure vs baryon density) for the neutrino-free case at $T = 30$ MeV. Clearly, the pressure becomes to close to the one given by the Maxwell construction at finite temperature, since the mixed phase becomes unstable and its density regime is thereby largely restricted. The mechanical instability of the pasta structure also arises in other geometrical shapes, which means that both R and R_W go to infinity with their ratio fixed. Thus we recover the picture of the phase equilibrium of two bulk matters, where the surface energy is irrelevant and the Coulomb energy vanishes due to the achievement of the local charge-neutrality. These features indicate that the EOS resembles the one given by the Maxwell construction.

2.8. Neutrino Trapping Effects

In Fig. 9, we show the dependence of the free energy on neutrino trapping in the mixed phase [20]. Here, we adopt the density-dependent bag, and fixed the quark volume fraction $(R/R_W)^3$ at $Y_{\nu_e} = 0.01$ for each curve. For $Y_{\nu_e} > 0.1$, the minimum point disappears. The right panel shows the each contribution to the free energy. From this figure, we can see that both the correlation energy and the Coulomb energy mainly contribute to the behavior of the minimum point in the presence of neutrinos, while it is a only the correlation energy for the temperature dependence.

To elucidate the neutrino-trapping effect in the mixed phase, we show the density profiles and the Coulomb potential for the slab case in Fig. 10. For

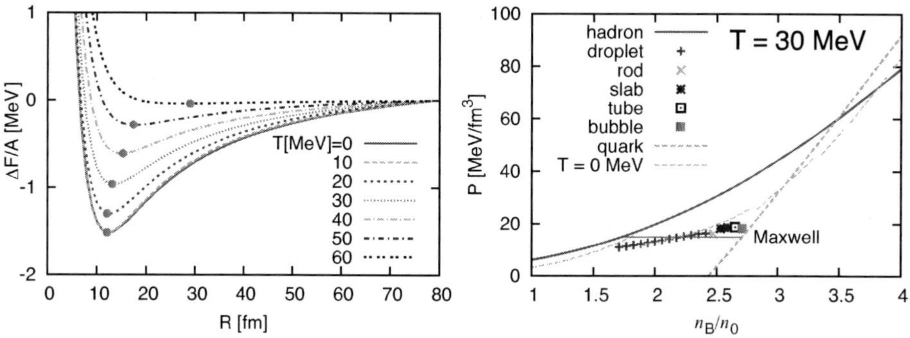

Figure 8. Left panel shows that the radius R dependence of the free-energy per baryon for the droplet case at $n_B = 2\ n_0$ and different temperatures. The free energy is normalized by the value at $R \to \infty$. The filled circles on each curve shows the energy minimum. The results are for $B = 100$ MeV·fm^{-3}, $\sigma = 40$ MeV·fm^{-2}. The right panel shows the EOS of the mixed phase (thick dots) in comparison with pure hadron and quark phases (thin curves) at $T = 30$ MeV. We also show, for comparison, the mixed phase by the Maxwell construction by thin solid line.

comparison, we use the same cell-size, and fix R_W here. High neutrino fraction enhances the number of electrons to satisfy the chemical equilibrium. As a result, the Coulomb potential drastically changes in the neutrino trapping case as shown in this figure.

In the left panel of Fig. 11, we show the relation of the pressure and density for PNS matter ($T = 30$MeV and $Y_L = 0.4$) as a result of above calculations. For comparison, we also check the phase transition under the Gibbs conditions with GCN and LCN. As shown in the left panel of Fig. 9, there is no minimum point of the free energy for PNS matter, which means the crystalline structures of pasta are broken there to form gan amorphous stateh composed of two species of matter; the amorphous state should take a bicontinuous structure with a complex pattern, depending on the kinetics of its formation. In such case, the surface tension and the Coulomb interaction do not work to make any pasta structure (see the right panel of Fig. 9), so that the mixed phase may resemble the amorphous phase of quarks and hadrons. For numerical reason, we can not calculate $R_W = \infty$. Instead, we set $R_W = 100$ fm for amorphous state.

From the left panel of Fig. 11 and the above discussion, the amorphous state of quarks and hadrons gives almost the same result with the bulk Gibbs calcula-

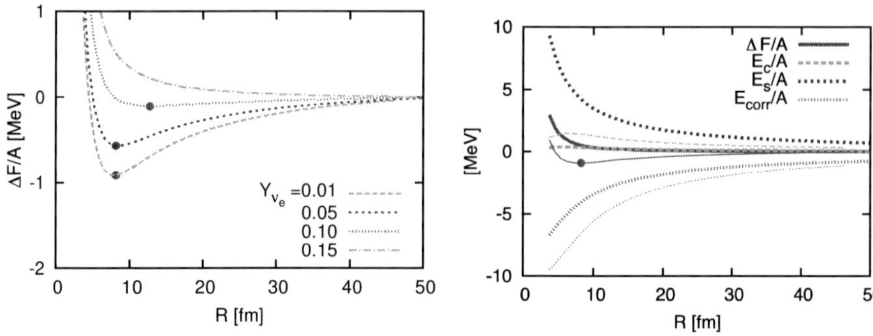

Figure 9. Left panel is the same as the left panel of Fig. 8, but for different neutrino fractions Y_{ν_e} at a given temperature, $T = 10$ MeV. In this case, we set baryon density as $n_B = 2.5\,n_0$. The thick lines of the right panel are the case of $Y_{\nu_e} = 0.15$, the thin lines $Y_{\nu_e} = 0.01$. The right panel shows each component of the free energy ($\Delta F/A$); e.g. the free energy, the Coulomb energy ($E_{\rm C}$/A), the surface energy ($E_{\rm S}$/A), and the correlation energy ($E_{\rm corr}$/A). Here they are shown per baryon.

tion with LCN. To understand this reason, we also show the density profile and the Coulomb potential $3n_0$ assuming the rod structure for the same PNS matter in the right panel of Fig. 11. From this figure, we can see that the shape of the Coulomb potential looks like a step function. This is because the size of R_W is very large ($R_W = 100$ fm); the surface effect becomes relatively small and the density profiles become almost uniform in each phase. Note again that the electron density is different in each case even if we impose the condition $\mu_e^H = \mu_e^Q$ in the presence of the Coulomb potential.

In the last of this subsection we summarize the above discussions for the HQ phase transition in Table 1. Basically the matter should be treated as a ternary system specified by the three kinds of the conserved quantities, baryon number, electric charge and lepton number. The proper treatment of the phase transition and the mixed phase proceeds by applying the Gibbs conditions. There are some kinds of variation instead of the proper treatment: the bulk Gibbs calculation (GCN) may be performed by neglecting the finite-size effects. The bulk Gibbs calculation has been also done with the use of the assumption of local charge neutrality (LCN). One may regard the matter as a binary system since

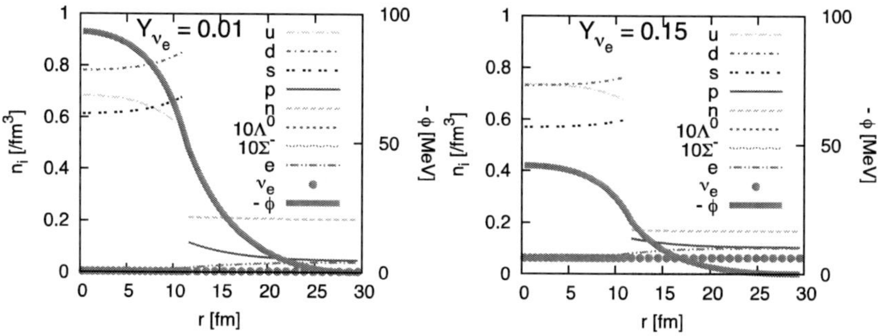

Figure 10. Density profiles and the Coulomb potential ϕ within 1D (slab) for $n_B = 2.5\ n_0$ at $T = 10$ MeV. Here, the neutrino fractions are set to $Y_{\nu_e} = 0.01$ (left panel) and $Y_{\nu_e} = 0.15$ (right panel). The cell sizes are $R_W = 30$ fm in these figures. The slab size are $R = 10.9$ fm. $(R/R_W)^3$ is fixed to be the optimal value at $Y_{\nu_e} = 0.01$.

the chemical equilibrium is required only for baryon and lepton numbers in this treatment. In the realistic treatment of the mixed phase, the finite-size effects are indispensable. When the finite-size effects are taken into account, we see the emergence of the inhomogeneous structure called pasta. We have seen that the proper treatment based on the Gibbs conditions gives almost the same EOS as the one given by the bulk Gibbs calculation with LCN in PNS matter. This is because the scale of inhomogeneity becomes large, e.g. $R_W \to \infty$, and an amorphous state is formed, where the finite-size effects are irrelevant and local charge neutrality is also attained. This is due to the mechanical instability by the neutrino trapping effect. Thus we can say the ternary system is effectively reduced to a"binary" system. We attach "binary" in the table to stress this situation.

The situation somewhat changes for various cases. Neutron star (NS) matter is a cold crystallized matter and includes no neutrino. So it should be treated as a binary system specified by the baryon number and electric charge. We have seen that it is effectively reduced to a "pure" system due to the mechanical instability for sufficiently large surface tension or high temperature; the scale of the inhomogeneity becomes large again and local charge neutrality is approximately achieved. Consequently the EOS resembles the one given by the Maxwell construction. The NS-NS mergers corresponds to the latter case, and we named "merger (MG)" matter. In this sense we listed the system as "pure"

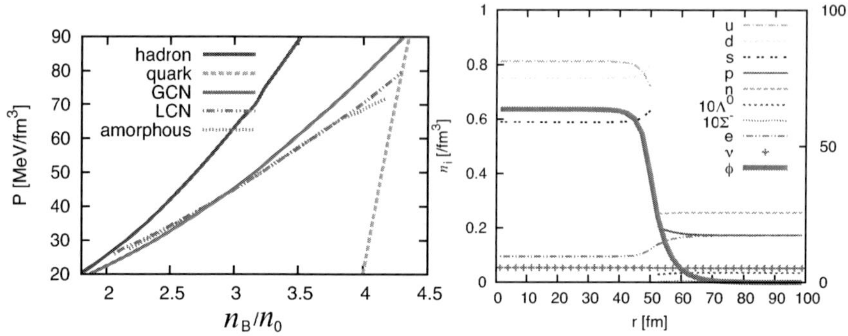

Figure 11. Left panel shows the pressure for PNS matter of $T = 30$MeV and $Y_L = 0.4$. The nuclear density is denoted by n_0. Right panel shows the density profile and the Coulomb potential $3n_0$ assuming the rod structure for the same PNS matter.

in the table for these cases. Note that there is one important difference between these cases and PNS matter, while the situation looks similar to each other; there is no amorphous state in these cases and appears a density jump inside compact stars instead.

3. Possibility to Probe the Properties of the Mixed Phase

As mentioned above, possible existence of the HQ mixed phase in high density region can be theoretically expected. However, it might be quite difficult on the Earth to reveal the properties of such an extreme state. The one of the possibilities to overcome this difficulty may be the direct observation of neutron stars. In fact, since the stellar structure of neutron stars strongly depends on their interior properties, one could obtain the information of the stellar interior via the observations of stellar mass M, radius R, and compactness defined as M/R.

Once the EOS of matter is obtained, we can extract the density profile (density as a function of radial distance from the center) and consequently the mass and the radius of compact stars. This is done by numerically solving the

Table 1. Comparison of conditions for the HQ phase transition

	finite-size effects	globally conserved variables	locally conserved variables	equilibrium conditions	system
Maxwell	No	n_B	Y_L, Y_C	$\mu_B^H = \mu_B^Q$	pure
bulk Gibbs (GCN)	No	n_B, Y_L, Y_C		$\mu_B^H = \mu_B^Q,\ \mu_L^H = \mu_L^Q,\ \mu_C^H = \mu_C^Q$	ternary
bulk Gibbs (LCN)	No	n_B, Y_L	Y_C	$\mu_B^H = \mu_B^Q,\ \mu_L^H = \mu_L^Q$	binary
pasta	Yes	n_B, Y_L, Y_C		$\mu_B^H = \mu_B^Q,\ \mu_L^H = \mu_L^Q,\ \mu_C^H = \mu_C^Q$	ternary
NS matter pasta	Yes	n_B, Y_C		$\mu_B^H = \mu_B^Q,\ \mu_C^H = \mu_C^Q$	binary
NS matter pasta (large σ)	Yes	n_B, Y_C		$\mu_B^H = \mu_B^Q,\ \mu_C^H = \mu_C^Q$	"pure"
MG matter pasta	Yes	n_B, Y_C		$\mu_B^H = \mu_B^Q,\ \mu_C^H = \mu_C^Q$	"pure"
PNS matter amorphous	Yes	n_B, Y_L, Y_C		$\mu_B^H = \mu_B^Q,\ \mu_L^H = \mu_L^Q,\ \mu_C^H = \mu_C^Q$	"binary"

Tolman-Oppenheimer-Volkoff (TOV) equation [49],

$$\frac{dp}{dr} = -\frac{Gm\epsilon}{r^2}\frac{(1+p/\epsilon)\left(1+4\pi r^3 p/m\right)}{1-2Gm/r}, \tag{25}$$

$$\frac{dm}{dr} = 4\pi r^2 \epsilon\,, \tag{26}$$

being G the gravitational constant. Starting with a central density $\epsilon(r=0) \equiv \epsilon_c$, one integrates out Eqs. (25) and (26) until the surface pressure becomes to zero. This gives the stellar radius R and its gravitational mass $M = m(R)$.

We show the effect of the HQ phase transition on the relation between the EOS and bulk structures of compact stars in Fig. 12. The left panel is the EOSs and the right panel shows the mass-radius relation of calculated neutron stars. As for the EOS, we shows three types of the HQ phase transitions; c.g. the EOS by the bulk Gibbs calculation, the Maxwell construction, and full calculation considering the finite-size effects. Here, the value of the surface tension is set as $\sigma = 40$ MeV$\cdot$fm^{-2}. We also show the uniform hadron and quark matter for comparison. Using them, we can calculate the mass-radius relations shown in the right panel. It is well-known that the typical masses of neutron stars are distributed in a narrow band close to 1.4 times the solar mass $M_\odot$ [50]. The maximum mass of neutron star is enough larger than these masses. With inclusion of hyperon degree of freedom, however, the softening of matter reduces the possible mass of neutron stars as low as $1.3\ M_{\odot}$, which contradicts with the most observations. The blue curves show the cases with the HQ mixed phase. The upper most one and the bottom are the results with the EOS obtained by the Maxwell construction and the bulk Gibbs calculation, respectively. The result with inclusion of the full pasta structures with $\sigma = 40\ \mathrm{MeV}\cdot\mathrm{fm}^{-2}$ lies just below the Maxwell construction case.

In any case with the HQ mixed phase, the calculated maximum mass is slightly above $1.4M_{\odot}$, which might be tolerable in confrontation with the observation. What is important is that the mixture of hyperons in nuclear matter softens the EOS, which causes the contradiction with the observation, i.e. the "maximum-mass problem". By considering the deconfinement transition to quark matter, the EOS gets stiffer at higher density region. This stiffness at high density sustains the neutron stars with $M \sim 1.4M_{\odot}$.

Recently Demorest et.al. have reported that there is a two-solar-mass neutron star through the measurement of the Shapiro delay [51]. Motivated by this observation, some authors studied the possibility of hard EOS considering exotic matter; c.g. Ref.[52] for hadron matter with hyperons, and Ref.[37] the matter including quarks. We are also planning to update our EOS models to other models to satisfy the observation.

Another possibility is the observations of gravitational waves radiated from neutron stars or the observations of stellar oscillations itself, where the observed spectra can reveal the interior properties [53, 54]. This technique is well-known as "gravitational-wave asteroseismology" or just "asteroseismology," which is similar to the seismology in the Earth and the helioseismology in the Sun. In this section, we focus on the possibility to see the properties of the HQ mixed phase via the observations of gravitational waves in subsection 3.1 and stellar oscillations in subsection 3.2.

In the last part of this section, we introduce another possibility to know the effects of the HQ mixed phase on compact-star phenomena; thermal cooling of neutron stars. A cooling theory of neutron stars has been well-discussed but a still uncertain issue [55, 56]. Cooling depends on the internal state which is in the range of ultrahigh density around/above the normal density. It may be directly connected to the EOS discussed in the previous section.

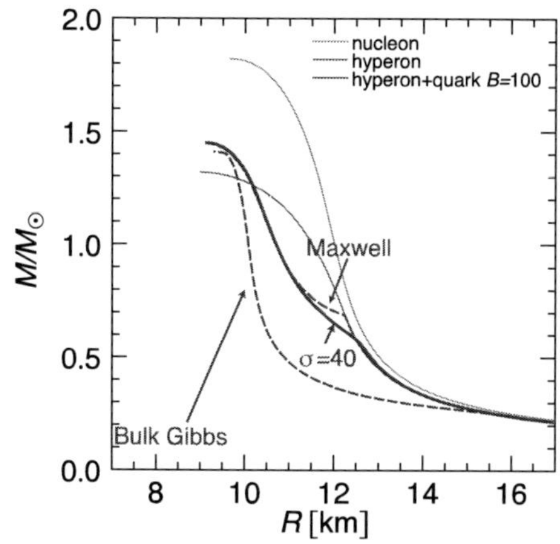

Figure 12. Mass-radius relation of calculated neutron stars. Green curve indicates the result with only nucleons.Red curve is that with hyperons. Blue curves are results with hadron matter and quark matter by different treatments of the mixed phase Note that the Maxwell construction is incorrect in the present case, but it is still useful as an eye guideline.

3.1. Gravitational Waves

The gravitational waves are oscillations of spacetime itself. Due to their strong permeability, the gravitational waves can carry away the raw information about the wave sources. So, observing the gravitational waves emitted from neutron stars, it could be possible to collect the astronomical data [53, 54, 57], to reveal the properties of dense matter [58, 59, 60, 61, 62, 63], and to probe the gravitational theory in the strong gravitational field [64, 65, 66, 67, 68]. Now, in order to directly detect the gravitational waves, several ground-based detectors are in operation, and the projects to build the next generation detectors are also moving ahead [69, 70, 71], which could enable us to observe the gravitational waves in the not-so-distant future.

Considering the gravitational waves radiated from the spherically symmetric stars, their oscillations can be classified into two families with their parities, i.e., odd (axial) and even (polar) parities. Since the odd parity oscillations are completely decoupled with the even parity oscillations, one can examine each type of oscillations separately. The odd parity oscillations are incompressible motion, while the even parity oscillations involve the deviation of density. Thus, from the observational point of view, the even parity oscillations might be important in the emission of gravitational waves. Here, in order to determine the specific oscillation frequencies in neutron stars, we adopt the relativistic Cowling approximation, where the perturbation of gravitational potential should be neglected. Namely, the fluid would oscillate on a fixed background spacetime. In practice, even with this simple approximation, one can qualitatively see the features for oscillation frequencies of emitted gravitational waves.

During the stellar oscillations, many kinds of gravitational waves could be radiated. The fundamental (f) and the pressure (p) modes can be excited in the spherically symmetric stars without the density discontinuity inside the star, while the gravity (g) modes can be also excited with the density discontinuity. As shown in Fig. 13, the stellar models with the EOSs adopted in this article except for the EOS with the Maxwell construction have only f and p modes, where the details how to determine the specific oscillation frequencies can be seen in Ref. [63]. On the other hand, in the stellar models constructed with the EOS adopted the Maxwell constructions whose masses are more than $0.7M_{\odot}$, the g mode is also excited as well as f and p modes. It should be noticed that since the stellar models with $M < 0.7M_{\odot}$ do not have density discontinuity inside the star because of the small central density, the g mode can not be excited. Since, as mentioned before, the g mode can be excited due to the existence of density discontinuity inside the star, the observations of the g mode can tell us such an information. In fact, the typical frequencies of g modes are in the range from a few hundreds Hz up to kHz, it could be possible to detect via the ground-based gravitational wave detectors. Additionally, as shown in Fig. 13, the frequencies of g mode for the stellar models adopted in this article are almost independent of the stellar mass, i.e., ~ 1.73 Hz. While, as shown in Fig. 14, it is found that the g mode frequencies can be well expressed as a function of stellar compactness, such as $\omega M = 0.3130(M/R) + 0.0103$. That is, via the detailed observations of g mode gravitational waves, one could know the stellar properties with the help of the additional observations about the stellar parameters such as the mass.

Furthermore, it is well-known that the f mode frequency can be associated with the stellar average density defined as $(M/R^3)^{1/2}$ [53, 54]. This could be physically explained by considering the relation between the sound speed and the propagation time of the fluid perturbation inside the star. The f mode frequencies calculated for the stellar models with the adopted EOSs are shown

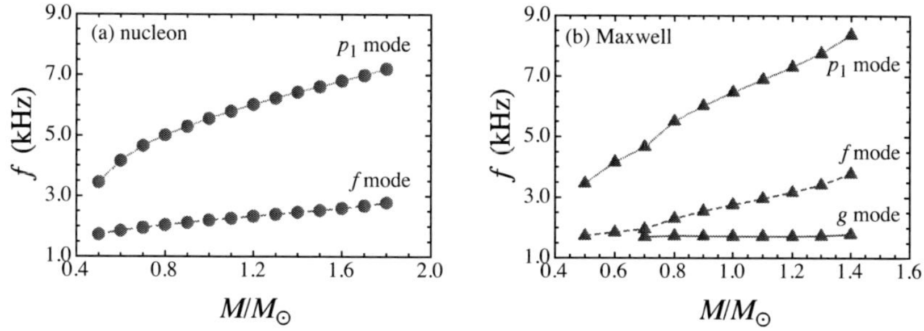

Figure 13. The frequencies of f and p_1 modes for the stellar models with (a) nucleon and (b) the HQ mixed phase adopted the Maxwell construction as a function of the stellar mass. In the right panel, the additional frequencies of g modes are also shown as well as the frequencies of f and p_1 [63].

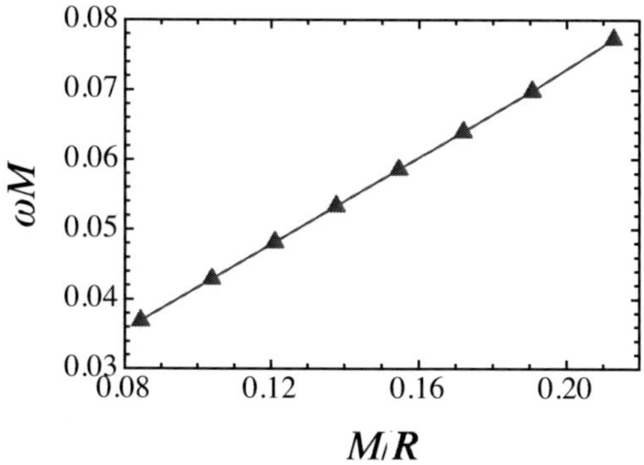

Figure 14. The normalized eigenfrequencies of g modes with EOSs considering the Maxwell construction as a function of the stellar compactness M/R [63].

in Fig. 15. Roughly speaking, one can observe from this figure that the frequencies for the stellar models without the density discontinuity, i.e., except for the EOS with the Maxwell construction, are almost similar behaviors. Thus, it might be difficult to reveal the properties of the HQ mixed phase via the observations of f mode gravitational waves. However, it should be emphasized that the f mode frequencies for the stellar model with the density discontinuity (with EOS adopted the Maxwell construction) can behave qualitatively differ-

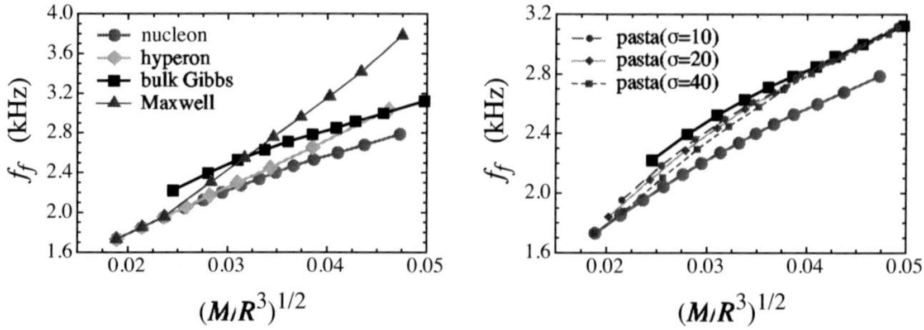

Figure 15. The frequencies of f modes for the stellar models with different EOSs as a function of the stellar average density $(M/R^3)^{1/2}$. The results with nucleon, hyperon, HQ mixed phase adopted the bulk Gibbs conditions (GCN) or the Maxwell construction are shown in the lift panel, while the right panel focuses on the results with the EOS considered the finite-size effects. Here the units of each σ are MeV $\cdot$ fm^{-2} [63].

ent from the others. That is, via the observations of not only g modes but also f modes gravitational waves, one might be able to prove the existence of the density discontinuity inside the star. Furthermore, from Fig. 16 for the normalized eigenfrequencies of f and p_1 modes, one can obviously see the difference between the frequencies of f mode gravitational waves with and without density discontinuity as mentioned the above, while the p_1 mode gravitational waves seem to depend on the adopted EOSs even if the density discontinuity does not exist inside the star. Therefore, the detailed observations of p mode gravitational waves could reveal the properties of the HQ mixed phase.

3.2. Shear Oscillations

Unlike the gravitational waves emitted from neutron stars, there exist the observed evidences of the stellar oscillations, i.e., the quasi-periodic oscillations (QPOs) in giant flares radiated from soft-gamma repeaters (SGRs). Up to now, three giant flares have been observed, such as SGR 0526-66 in 1979, SGR 1900+14 in 1998, and SGR 1806-20 in 2004, and the existence of a lot of QPO frequencies in the afterglow of giant flares are found via timing analysis, whose frequencies are in the range of tens Hz up to kHz [72]. In order to theoretically explain the observed QPO frequencies, many examinations have

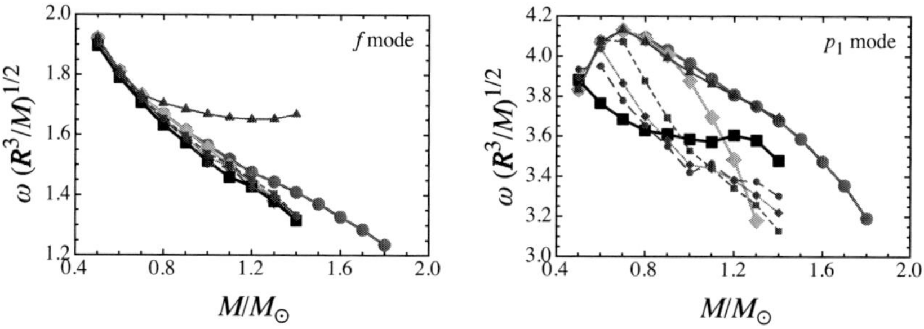

Figure 16. The normalized eigenfrequencies of f modes (left panel) and p_1 mode (right panel) as a function of the stellar masses. The marks in the figures are the same meaning as in Fig. 15 [63].

been done in terms of the torsional oscillations in the crustal region and/or the magnetic oscillations (e.g., [73, 74, 75, 76, 77, 78, 79, 80, 81]). If the observed QPO frequencies are associated with the torsional oscillations in neutron star crust, one might be able to obtain the information about the crust properties [82, 83, 84, 85, 86]. The torsional oscillations are incompressible, i.e., odd parity oscillations, where the elasticity plays a role as a restoring force, and such elasticity can be characterized by the shear modulus μ [87]. Although the derivation of μ for the complex structure have never been done, the expression of μ due to the Coulomb interaction in the body center cubic (bcc) in neutron star crust is known as

$$\mu = 0.1194 n_i (Ze)^2/a, \tag{27}$$

where n_i, Ze, and a are ion number density, charge of ion, and average ion spacing defined as $4\pi a^3/3 = 1/n_i$, respectively [88]. Note that, averaging the all directions and assuming the zero temperature, this expression is derived from the Monte Carlo calculations.

On the other hand, as mentioned the above, the possibility to appear the nonuniform structure (the HQ pasta phase) in the high density region inside the neutron stars is suggested [18, 19], which is similar structure to the pasta phase in the neutron star crust [7, 8]. If this suggestion is true, similar to the neutron star crust, one might be able to consider the shear in the HQ mixed phase, and the torsional oscillations could be excited even in the HQ mixed phase [89, 90]. In order to examine such torsional oscillations, for simplicity, we consider the shear only for the quark spherical droplet in the uniform hadronic matter, although the different nonuniform structures could appear in the HQ pasta phase, such as cylindrical and slab structures. Due to this assumption, the shear in the HQ mixed phase effectively makes an underestimation. So, the calculated frequencies of torsional oscillations could be corresponding to the lower limit, because the frequencies of torsional oscillations can be proportional to the shear speed defined as $v_s = (\mu/\rho)^{1/2}$. Moreover, to adopt the expression for μ, Eq. (27), we consider that n_i should be the number density of quark spherical droplet in the hadronic sea, while Ze should be the total charge included in the quark spherical droplet. As a typical value of the surface tension σ, we adopt three values, such as $\sigma = 10$, 20, and 40 MeV $\cdot$ fm^{-2}. The detail for calculations can be seen in [90].

The calculated frequencies of the $\ell = 2$ fundamental torsional oscillations in the HQ mixed phase, ${}_0t_2$, are shown in Fig. 17 as a function of the stellar mass. Since the frequencies of fundamental torsional oscillations in the neutron star crust are around tens Hz [76, 87], one can observe around 10 times larger frequencies in the HQ mixed phase. This is why the shear modulus in the mixed phase calculated from Eq. (27) becomes about 10^3 times larger than that in crust. As a result, the propagation time with the shear speed becomes about 10 times smaller than that in the crust. Then, one can estimate that the frequencies of torsional oscillations in the mixed phase could become roughly 10 times as large as those in crust region. Additionally, one can see from Fig. 17 that the frequencies of fundamental torsional oscillations depend strongly on the value of σ. In practice, the frequencies for $\sigma = 20$ and 40 MeV$\cdot$ fm^{-2} are $\sim$40% and $\sim$120 % larger than those for $\sigma = 10$ MeV$\cdot$ fm^{-2}. Thus, if one would identify the observed frequencies as the torsional oscillations in the HQ mixed phase, one can probe the properties of such exotic structure. Additionally, with the fixed stellar mass, we plot the frequencies of fundamental torsional oscillation with $\ell = 2$ in the HQ mixed phase as a function of surface tension σ in Fig. 18. From this figure, one can obviously observe the linear relation between the frequencies and surface tension. Thus, with the help of the other observation of stellar mass, one can make a constraint in the value of σ via the observation of the frequencies of fundamental torsional oscillations in the HQ mixed phase. Now, we should remark the possibility of the detection of this type oscillation. In fact, if the star is exactly spherically symmetric, one might have no chance to observe this oscillation, because the HQ mixed phase is surrounded by the hadronic fluid matter and the oscillation is completely con-

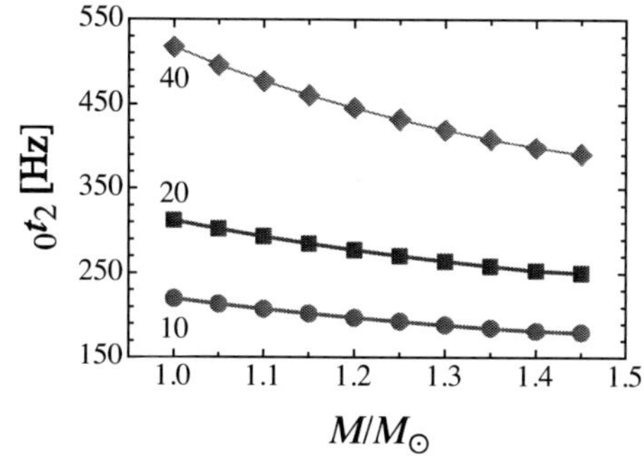

Figure 17. With $\sigma = 10$, 20, and 40 MeV$\cdot$ fm^{-2}, the frequencies of the $\ell = 2$ fundamental torsional oscillations ${}_0t_2$ are shown as a function of stellar mass $M/M_\odot$ [90].

fined in the HQ mixed phase. But, considering the realistic stellar model, there exist the rotational and/or magnetic effects, which make a deformation of stellar shape. Under this situation, the torsional oscillations with axial parity can be coupled with the even parity oscillations mentioned in Sec. 3.1. Probably, this is one opportunity to observe the torsional oscillation in the HQ mixed phase. Anyway, the examinations in the more complicated system will be done somewhere. Furthermore, since the resulting frequencies of fundamental torsional oscillations are order of 100 Hz, some of the QPO frequencies observed in giant flares, for example 150 Hz and even 626.5 Hz in SGR 1806-20 and 155 Hz in SGR 1900+14 [72], might be associated with the torsional oscillations in the HQ mixed phase.

3.3. Cooling of Neutron Stars with the Hadron-Quark Mixed Phase

Now we focus on the cooling theory of isolated neutron stars. Isolated neutron stars do not have internal heat source (someone says magnet-breaking heating affects to the cooling, but the effect is not so large), and they are born at the sites of supernovae explosion where the temperature rises to 30-50 MeV in the core of PNS as we described in the last section. Isolated neutron stars evolve thermally by emitting energy and losing their temperature immediately, as a result the temperature becomes to a few MeV (or $\sim 10^{10}$K) in a minute[91, 92]. In this subsection we discuss the cooling of neutron stars after a few minutes

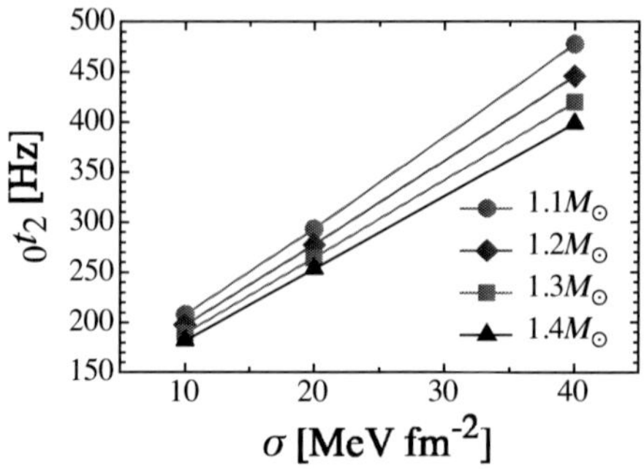

Figure 18. Frequencies of fundamental torsional oscillations with $\ell = 2$ as a function of surface tension with different values of stellar mass [90].

from their birth. Hence initial temperature is assumed as ~ 1 MeV in our calculations.

The most efficient energy-release process in neutron stars is the neutrino emission, based on neutron beta-decay. The well-known process named direct URCA process is combination of neutron beta-decay and inverse reaction.

$$n \rightarrow p + e^- + \bar{\nu}_e, \quad p + e^- \rightarrow n + \nu_e \tag{28}$$

Working forward and inverse (electron capture) process, electron neutrinos and anti-electron neutrinos are emitted and carry some energy. Emitted neutrinos travel through the stars without any interactions due to the large mean free path. However, due to the abundances of proton and neutron, direct URCA process is kinematically forbidden in the neutron stars [49]. To solve this issue, the modified URCA process was promoted, which involves another neutron to conserve momentum.

$$n + n \rightarrow n + p + e^- + \bar{\nu}_e, \quad n + p + e^- \rightarrow n + n + \nu_e \tag{29}$$

The modified URCA process has a lower emissivity; it is about five orders of magnitude weaker than one with the direct URCA.

Other processes also work on neutron stars cooling, such as Bremsstrahlung (neutrino emission) and photon emission, but they work dominantly in the crust (Bremsstrahlung), or the surface (photon emission) of neutron stars, and their emissivities are weaker than the core processes.

Modified URCA and other processes could explain most of neutron stars observational data, with some exception. Therefore the "Exotic (Non-Standard) Cooling" which works in some kinds of exotic phases, is introduced and the ordinary cooling is called the "Standard Cooling" [93]. The standard cooling works in all the neutron stars, and the exotic cooling works in some of neutron stars. The exotic cooling works above the threshold density, due to some kind of exotic situations where the exotic phase appears. The exotic situation can be said as pion condensation, kaon condensation, existence of hyperons or quark matter [49]. In the exotic state, the neutrino emission process plays as the direct URCA, due to the degrees of freedom. Once the exotic cooling process works, star cools faster than standard one (eg. [94, 95]).

Stars with the exotic cooling have larger central density than ones with the standard cooling. The mass of stable neutron stars increases with the central density, therefore stars with the exotic cooling have larger mass than standard ones [96]. Also, the neutrino emissivities of these processes including the modified URCA have density dependent; it is monotonically increasing with the density. It is believed that lighter neutron stars cools slower and heavier one cools faster.

The matter in neutron stars is well degenerated; neutrons and protons can create the Cooper pairs and condensate to superfluid (superconducting for protons) state at some temperature. The transition between normal to superfluid state, the latent heat escapes by causing neutrino pair creation [97]. This process is called as PBF (Pair Breaking and Formation) process, and it works only where the temperature of the region is near to the critical temperature of superfluidity. In neutron stars, there are three possible nucleon superfluid states ($^1S_0$ and $^3P_2$ for neutrons, and $^1S_0$ for protons), and they might have three PBF working regions. The neutron $^3P_2$ state works at the highest density of these three, and it may become the dominant cooling process at specific density and time [98, 99]. The transition does not last long time and occurs just once in a density region, therefore the integrated effect is limited. However, once matter becomes to superfluid state, thermodynamical properties decreases and the emissivity of neutrino also decreased. Nucleon superfluidity works both of acceleration and breaking of cooling. The neutrino emission by the transition to the superfluid state works all the neutron stars, and called as "minimal cooling" including the modified URCA, Bremsstrahlung and PBF [100].

There are some cooled stars whose effective temperature cannot be explained by the standard cooling scenario, which does not include the neutrino emission caused by nucleon superfluidity. It needs stronger cooling process as in the case of J0205+6449 in 3C58 (hereafter, "3C58") or Vela pulsar (B0833-45). Also an accreting neutron star SAX J1808 requires strong cooling. 3C58 and Vela may be explained by the minimal cooling model which includes nucleon superfluidity [101]. However, SAX J1808 needs stronger cooling than the minimal cooling [102].

In recent observation, two important data for cooling of neutron stars are observed and analyzed. One is the central source of Cassiopeia A (hereafter "Cas A") [103, 104] and the other is J1614-2230 [51]. They give strong constraints on the cooling of neutron stars. J1614-2230 is discussed in the beginning of this chapter, we introduce about the Cas A issue here.

Cas A is the youngest-known supernova remnant in our Milky Way and it locates ~ 3.4 kpc distant from the solar system [105]. The supernova explosion occurred about 330 years ago, but due to absorption by interstellar medium, there are no exact historical records except for unclear detection by J. Flamsteed in 1680 [106]. Recently, [103] and [104] have analyzed the X-ray spectra of Cas A. They give the effective temperature and possible regions occupied by the mass-radius relations. Since Cas A is the isolated remnant, uncertainty of the mass-radius relation could be large. The lowest mass obtained from χ^2 fitting is about $1.5M_\odot$. Considering the age of $t = 330$ yr, the effective temperature $T_{\rm eff}$ of Cas A must occupy a point of a cooling curve due to the standard cooling scenario on the $(T_{\rm eff} - t)$ plane. This gives strong constraint on the EOS and cooling processes. Furthermore, [104] reported the observation of $T_{\rm eff}$ for Cas A in recent 10 years. [107], [108] and [109] insist that the rapid decrease in $T_{\rm eff}$ during the years shows that the transition to nucleon superfluidity occurs.

3.3.1. Cooling of Hybrid Stars with the Hadron-Quark Mixed Phase

Considering the HQ mixed phase for the EOS of neutron star, it is important to apply for the cooling theory. The quark matter is known as an exotic matter which causes strong neutrino emission. Therefore, once the quark phase appears, strong neutrino emission cools the star quickly. The neutrino emissivity is calculated as [110]

$$\varepsilon_{\nu,q} = 8.8 \times 10^{26} \alpha_s \frac{n_B}{n_0} Y_e^{1/3} T_9^6 \ \mathrm{erg\,s^{-1}\,cm^{-3}}. \tag{30}$$

This emissivity can work inside of quark matter, and we show the following assumption.

We constructed a model which includes the HQ mixed phase. Considering the first order phase transition between the hadron and quark phases, it would be plausible that both phases coexist and form some kinds of the mixed phase. It has been shown that the mixed phase could form geometrical structures [18]; [19] have made EOS of mixed-phase under a Wigner-Seitz (hereafter "WS") approximation using the MIT Bag model for the quark phase at finite temperature.

We employ an EOS with the same framework using the bag constant $B = 100$ MeV fm^{-3}, the coupling constant $\alpha_S = 0.2$, and the surface tension parameter $\sigma = 40$ MeV fm^{-2}. For the hadron phase, we adopt results of the Bruekner-Hartree-Fock (BHF) theory including hyperons, Λ, and Σ^- [23, 24, 25, 26]. Since the BHF results is inappropriate for low-density matter in the crust, we apply EOS of BPS [111] for the crust. The EOS gives the maximum mass $1.53 M_\odot$ with the radius 8.6 km, and the mass lies within the limits of the observation of Cas A.

Using the WS approximation, we obtain a cell radius of each phase, and calculate the volume fraction of quark matter in the mixed phase as seen in Fig. 19.

It is difficult to calculate the neutrino emissivity in the mixed phase directly. Therefore, the volume fraction F is multiplied to the original quark neutrino emissivity $\varepsilon_{\nu,q}$ of (30); The total emissivity by quarks is set to be $\varepsilon_\nu = F\varepsilon_{\nu,q}$. We adopt the well-known standard cooling process for hadronic matter [93]: Modified URCA process for the higher density region and Bremsstrahlung process for the crust.

Considering the HQ mixed phase, it becomes possible to reduce the neutrino emissivity by quarks. However, the cooling rate is still too strong to explain cooled stars with lower limit (such as Vela), it requires some kind of reduction of neutrino emissivity.

3.3.2. Cooling of Hybrid Stars with the Hadron-Quark Mixed Phase and Color Superconductivity

Fortunately, there is possible factor of reduction of neutrino emissivity in the quark phase. It is the color superconducting (hereafter "CSC") phase of quarks [112]. There are some sorts of quark pairings such as CFL or 2SC according to the degrees of freedom of quark flavor and color. In the CFL phase, all colors and flavors quarks have a common Fermi momenta, and all quarks create pairs and condensated. Meanwhile u and d quarks of two colors have same

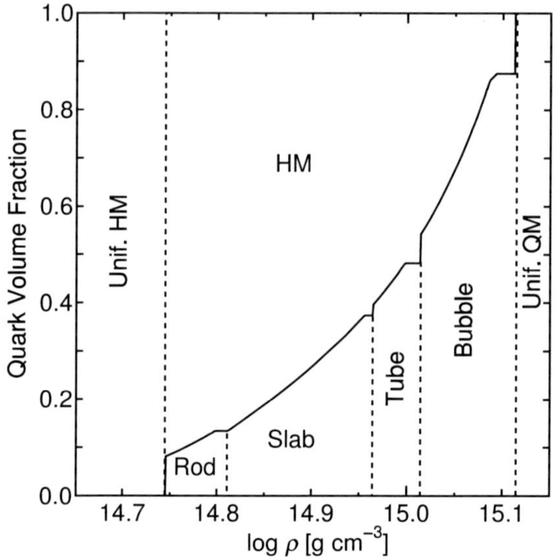

Figure 19. Volume fractions of quark matter phase having particular geometrical structures with a bag constant $B = 100$ MeV fm^{-3} and a coupling constant $\alpha_s = 0.2$ [18, 19].

Fermi momenta in the 2SC phase, and some quarks are unable to create pairs [3]. The CFL phase is likely to appear at higher density than the 2SC phase [113]. It is considered that the energy gap of $\Delta \sim 10$ MeV is very large compared with the temperature of the center of compact stars , $T_C \sim$ keV,(eg. [114]). Once matter becomes superconducting, neutrino emissivity must be suppressed similar to the hadronic superfluidity [115]. Due to the large energy gap and it could be proportional to $\exp(-\Delta/k_B T)$, where T is the temperature at the relevant layer and k_B is the Boltzmann constant. Therefore, in the color superconducting phase with a large energy gap ($\Delta \gg T$), neutrino emissivity by quarks is almost negligible [3]. We note that we do not need to consider which kinds of CSC pairing appears, if we adopt a large enough energy gap.

We assume that the energy gap in the color superconducting phase is very large and the phase appears in the density region above the threshold density ρ_T. For simplicity, we use the critical volume fraction of the quark matter in the HQ mixed phase F_C instead of ρ_T. If the matter has a volume fraction

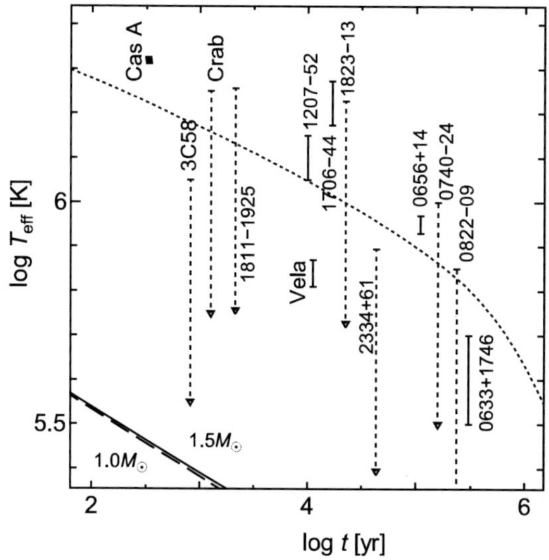

Figure 20. Cooling curves with the HQ mixed phase. Solid and dashed curves denote cooling curves of the hybrid stars with the HQ mixed phase, and the dotted curve denotes the cooling curve of the neutron star with standard cooling. The curves show the models of hybrid stars with the HQ mixed phase become too cold to explain observed data.

$F > F_C$ at the density in the layer of the mixed phase, quarks become the color superconducting state.

We select stellar masses of compact stars to be 1.0, 1.3, and $1.5M_\odot$ whose central densities are 1.48, 1.82, and 2.67 $\times 10^{15}$ g cm^{-3}, respectively. We take the critical fraction F_C to be 0.1, 0.125, and 0.2 which are appropriate to explain the observations with our scenario. Results are shown in Fig. 21, with available observational values. Most of the observational data are taken from the Table 7.2 in [116] : [117] for 3C58, [118] for Crab, [119] for Vela, [120] for 0656+14, [121] for 1706-44, [116] for 1811-1925, [122] for 1823-13, [123] for 2334+61. Other data are taken from [124]. The data of Cas A is attached [104] as the most youngest compact star. The cooling curves transit from hotter region to cooler region for the parameter between $0.1 < F_C < 0.2$. As indicated in the middle panel of Fig. 21, cooling curves split into two regions for $F_C = 0.125$

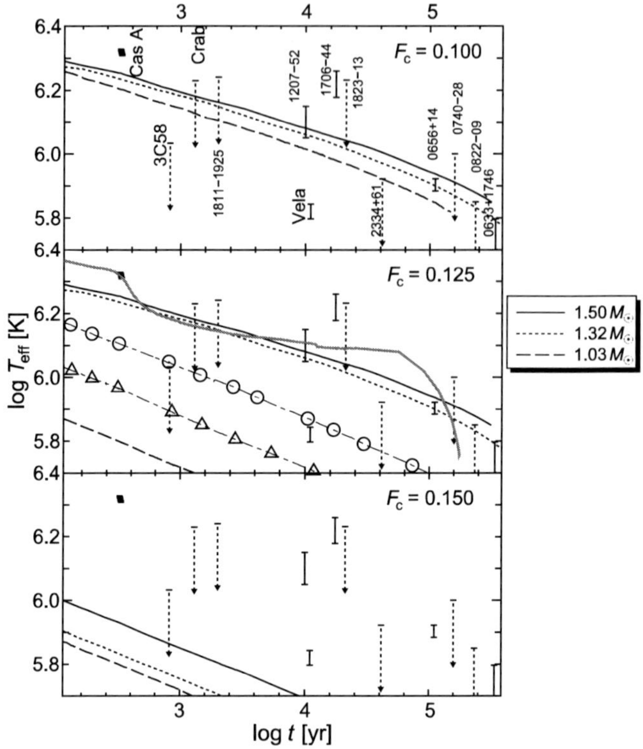

Figure 21. Cooling curves with color superconducting quark phases [125]. Solid, dotted, and dashed lines denote the models with the masses of 1.50, 1.32, and 1.03 $M_\odot$, respectively. Thick gray line on the middle panel denotes a $1.50M_\odot$ model with PBF and carbon envelope. Dot-dashed lines with marks on the middle panel indicate the model of the mass $1.03M_\odot$ except for the neutrino emissivity in normal quark phase multiplied by $1/10$ and $1/100$ for the lines with marks of circle and triangle, respectively.

and stars with larger masses cool slower than those with lighter masses. Since lighter mass stars cool faster, they suit for the 3C58 case who does not have lower limit of the effective temperature observation. However, calculated cooling curves are inconsistent with the Vela case which has lower limit and the quark cooling is still too strong to explain this case.

Since the neutrino emissivity of the quark phase involves large uncertainty, we have calculated the additional cooling curves for the mass $1.03M_\odot$ in case of $F_C = 0.125$ with the neutrino emissivity reduced by a factor of 0.1 and 0.01. There are some possible factors of this reduction for neutrino emissivity accompanying quark β-decay, such as increase of the abundance of strange quarks; decreasing in electron numbers inside MP, leads to the reduction of the neutrino emissivity [110]. The presence of 2SC at low density also suppress the emissivity; [18] discussed that the abundance of quarks in the mixed phase changes and it may cause the color superconducting phase. We suppose that the reduction of the emissivity originates from the above physical processes. If the emissivity of quarks is reduced by these factors, the observation of Vela can be explained as shown in the middle panel of Fig. 21. Also, we calculated the effect of PBF and the carbon envelope with our model to fit the recent Cas A data [104], and the result is shown in the thick gray line in the middle panel of Fig. 21.

Considering the HQ mixed phase, the cooling theory of neutron star requires the existence of some kind of color superconducting phase with large gap. If the mass of an isolated neutron star other than Cas A is detected in near future, it might help to probe the existence of color superconducting phase, quark matter and the HQ mixed phase.

4. Summary and Concluding Remarks

We have discussed features of the HQ mixed phase in compact stars, where multi components are in chemical equilibrium. Such phase transitions are generally not congruent and the Maxwell construction, which is a common method to obtain the equation of state (EOS) of matter with a single component, cannot be applied any more and one has to solve the Gibbs conditions instead. Using the bulk calculation, i.e., without considering the surface tension and the Coulomb interaction among two phases, we have presented basic concepts of the non-congruent transition. Although these concepts are helpful in discussing and analyzing the properties of the phase transitions in compact stars, we must bear in mind that they are obtained in an idealized and simplified situation. For realistic cases, we have to take into account rather complicated finite-size effects such as the surface tension and the Coulomb repulsion.

When the finite-size effects are present, crystalline structures of regular geometries called "pasta structures" emerge in the mixed phase. Charge screening, i.e., the rearrangement of charged particle densities in the presence of the

Coulomb interaction is also emphasized; it sometimes causes an instability of the geometrical structures in the mixed phase. In the extreme case the EOS resembles the one given by the Maxwell construction, and uniform nuclear matter can exist in the mechanically unstable region.

We note that the maximum mass of hybrid stars in our calculation is rather low in the light of a recent observation [51], since contamination of hyperons considerably softens EOS. We have found that the transition of hadronic matter to quark matter suppresses the appearance of hyperons, but the maximum mass is still low due to the rather soft EOS of quark matter. Therefore, we need other EOSs of hadron and quark matter to circumvent this situation.

So far, most works on the pasta structures have used the WS cell with ansatz about the geometrical structures like droplet, rod, slab and so on. If we want to see the possibility of the intermediate shapes, or study the mixed phase in a more realistic way, Frameworks without the WS approximation or ansatz on the geometrical structures are desirable. Recently we have performed a three-dimensional calculation of non-uniform nuclear matter based on the relativistic mean-field model and the Thomas-Fermi approximation [12]. Introducing a cubic cell with periodic boundary conditions and discretizing it into grids, we numerically solve the coupled field equations for meson mean-fields and the Coulomb potential. Randomly distributing fermions (n, p, e) on the grid points as an initial condition, we relax their density distributions to attain the uniformity of the chemical potentials. We have shown that typical pasta structures, obtained in the previous studies with the WS approximation, also appear in the three-dimensional calculation [12].

Possibilities of observing the signals of the mixed phase has been suggested in the spectra of the gravitational waves [63]. We have found that one could know the existence of density discontinuity inside the star via observing the gravitational waves of not only g mode but also f mode. The detailed observations of p mode gravitational waves could reveal the properties of the HQ mixed phase because the p_1 mode gravitational waves depend on the adopted EOSs even if the density discontinuity does not exist.

QPO in giant flares from SGR is another probe from neutron stars. If the observed QPO frequencies are associated with the torsional oscillations in neutron star crust, one might be able to obtain the information about the crust properties. In fact, we have pointed out that some of the QPO frequencies observed in giant flares might be associated with the torsional oscillations in the HQ mixed phase.

Phenomenological implications of pasta structures should deserve more elaborate studies of their thermodynamical or dynamical properties, e.g., the neutrino opacity or the thermal conductivity may be affected by the pasta structures, and their elasticity may cause a discontinuous change of internal structure of compact stars. In particular, the crystalline structure of the HQ mixed phase in the core region may bring about novel dynamical effects on compact-star phenomena by way of deformation or oscillation.

The other important application of the EOS with the HQ mixed phase is the cooling theory of neutron stars, the long time issue discussed for decades. The cooling of neutron stars strongly depends on the neutrino emissivity of the core of stars. Simply applying the HQ mixed phase with normal quark matter, stars cool rapidly and do not match observations. However, including color superconductivity, the neutrino emissivity of the quark phase is much suppressed, and the cooling of stars with the HQ mixed phase does not conflict with observations.

Taking account of a color superconductivity in the HQ mixed phase, it became possible to have heavier stars cooling slower and lighter ones faster. We cannot get this situation without considering the HQ mixed phase. If one observes the masses of isolated neutron stars and the mass–temperature distribution can be described by the above scenario, it proves the existence of the color superconductivity in the HQ mixed phase.

Acknowledgments

NY is grateful to H.-J. Schulze, F. G. Burgio, and M. Baldo for their warm hospitality and fruitful discussions. This work was partially supported by the Grant-in-Aid for the Global COE Program "The Next Generation of Physics, Spun from Universality and Emergence" and for Scientific Research on Innovative Areas from the Ministry of Education, Culture, Sports, Science and Technology (MEXT) of Japan and the Grant-in-Aid for Scientific Research (C) and for Young Scientists (B) from JSPS (16540246, 20540267, 23105711, 24740177).

References

[1] P. Petreczky, arXiv:1203.5320.

[2] T. Schaefer, arXiv:0509068;
P. Braun-Munzinger and J. Wambach, *Rev. Mod. Phys.* 81 (2009) 1031.

[3] M. G. Alford, A. Schmitt, K. Rajagopal, and T. Schäfer, *Rev. Mod. Phys.*, 80 (2008) 1455.

[4] K. Fukushima and T. Hatsuda, *Rept. Prog. Phys.* 74 (2011) 014001.

[5] N.K. Glendenning, "*Compact Stars*", Springer, 2000.

[6] N.K. Glendenning, *Phys. Rev.* D46 (1992) 1274.
N.K. Glendenning, *Phys. Rep.* 342 (2001) 393.

[7] C. P. Lorenz, D. G. Ravenhall, and C. J. Pethick, *Phys. Rev. Lett.* 70 (1993) 379.

[8] K. Oyamatsu, *Nucl. Phys.* A561 (1993) 431.

[9] K. Sumiyoshi, K. Oyamatsu, and H. Toki, *Nucl. Phys.* A595 (1995) 327.

[10] T. Maruyama, T. Tatsumi, D.N. Voskresensky, T. Tanigawa and S.Chiba, *Phys. Rev.* C72 (2005) 015802.

[11] W.G. Newton and J.H. Stone, *Phys. Rev.* C79 (2009) 055801.

[12] M. Okamoto, T. Maruyama, K. Yabana and T. Tatsumi, *Phys. Lett.* B713 (2012) 284.

[13] H. Heiselberg, C.J. Pethick and E.F. Staubo, *Phys. Rev. Lett.* 70 (1993) 1355.

[14] D. G. Ravenhall, C. J. Pethick and J. R. Wilson, *Phys. Rev. Lett.* 27 (1983) 2066.

[15] D. N. Voskresensky, M. Yasuhira and T. Tatsumi, *Phys. Lett.* B541 (2002) 93; *Nucl. Phys.* A723 (2003) 291.

[16] T. Endo, T. Maruyama, S. Chiba, and T. Tatsumi, *Nucl. Phys.* A749 (2005) 333; *Prog. Theor. Phys.* 115 (2006) 337.

[17] T. Maruyama, T. Tatsumi, V. N. Voskresensky, T. Tanigawa, and S. Chiba, *Phys, Rev.* C73 (2006) 035802.

[18] T. Maruyama, S. Chiba, H.-J. Schulze and T. Tatsumi, *Phys. Rev.* D76 (2007) 123015; *Phys. Lett.* B659 (2008) 192.

[19] N. Yasutake,T. Maruyama, T. Tatsumi, *Phys. Rev.* D80 (2009) 123009.

[20] N. Yasutake,T. Maruyama, T. Tatsumi, arXiv:1202.0143.

[21] M. Alford, K. Rajagopal, S. Reddy and F. Wilczek, *Phys.Rev.* D64 (2001) 074017.

[22] H. Nemura, N. Ishii, S. Aoki, and T. Hatsuda, *Phys. Lett.* B673 (2009) 2.

[23] M. Baldo, "*Nuclear Methods and the Nuclear Equation of State*", World Scientific, Singapore, (1999).

[24] M. Baldo, G. F. Burgio, and H.-J. Schulze, *Phys. Rev.* C58 (1998) 3688.

[25] H.-J. Schulze *et al.*, *Phys. Lett.* B355 (1995) 21.

[26] I. Vidaña *et al.*, *Phys. Rev.* C61 (2000) 025802.

[27] R. B. Wiringa, V. G. J. Stoks, and R. Schiavilla, *Phys. Rev.* C51, (1995) 38.

[28] N.K. Glendenning and S. Pei, *Phys. Rev.* C52 (1995) 2250.

[29] B. S. Pudliner *et al.*, *Phys. Rev.* C56 (1997) 1720.

[30] P. M. M. Maessen, T. A. Rijken, and J. J. de Swart, *Phys.Rev.* C40 (1989) 2226.

[31] M. Baldo and L. S. Ferreira, *Phys. Rev.* C59 (1999) 682.

[32] O. E. Nicotra, M. Baldo, G. F. Burgio, and H.-J. Schulze, *Astron. Astrophys.* 451 (2006) 213.

[33] O. E. Nicotra, M. Baldo, G. F. Burgio, and H.-J. Schulze, *Phys. Rev.* D74 (2006) 123001.

[34] M. Baldo, L. S. Ferreira, and O. E. Nicotra, *Phys. Rev.* C69, (2004) 034321.

[35] G. F. Burgio, S. Plumari, *Phys. Rev.* D77 (2008) 085022.

[36] N. Yasutake, K. Kashiwa, *Phys. Rev.* D79 (2009) 043012.

[37] K. Masuda, T. Hatsuda and T. Takatsuka, arXiv:1205.3621.

[38] K. Fukushima, *Phys. Rev.* D77, (2008) 114028.

[39] K. Fukushima and H.J. Warringa, *Phys. Rev.* D78, (2008) 039902.

[40] K. Kashiwa, M. Yahiro, H. Kouno, M. Matsuzaki, Y. Sakai., *J. Phys. G. Nucl. Part. Phys.* 36, (2009) 105001.

[41] H. Kouno, Y. Sakai, T. Sasaki, K. Kashiwa, M. Yahiro, *Phys. Rev.* D82, (2010) 076003.

[42] O. Lourenço, M. Dutra, A. Delfino, M. Malheiro, *Phys. Rev.* D84, (2011) 125034.

[43] O. Lourenço, M. Dutra, A. Delfino, M. Malheiro, *Phys. Rev.* D85, (2012) 097504.

[44] H. Chen, W. Yuan, L. Chang, Y-X. Liu,T. Klähn, C. D. Roberts, *Phys. Rev.* D78 (2009) 116015.

[45] G. Pagliara, M. Hempel and J. Schaffner-Bielich, *Phys. Rev. Lett.* 103 (2009) 171102.
M. Hempel, G. Pagliara, J. Schaffner-Bielich, *Phys. Rev.* D80 (2009) 125014.

[46] For a review, T. Maruyama,T. Tatsumi,T. Endo and S. Chiba, *Recent Res. Devel. Phys.* 7 (2006) 1.

[47] E. Farhi and R. L. Jaffe, *Phys. Rev.* D30, (1984) 2379.

[48] K. Kajantie, L. Kařkaïnen, and K. Rummukainen, *Nucl. Phys.* B357 (1991) 693.

[49] S. L. Shapiro and S. A. Teukolsky, "*Black Holes, White Dwarfs, and Neutron Stars*", John Wiley & Sons Inc, 1983.

[50] For a recent review, see D. Page and S. Reddy, *Annu. Rev. Nucl. Part. Sci.* 56 (2006) 327.

[51] P. B. Demorest, T. Pennucci, S. M. Ransom, M. S. E. Roberts, and J. W. T. Hessels, *Nature,* 467 (2010) 1081.

[52] S. Weissenborn, D. Chatterjee and J. Schaffner-Bielich, *Phys. Rev.* C85 (2012) 065802.

[53] N. Andersson and K. D. Kokkotas, *Phys. Rev. Lett.* 77 (1996) 4134.

[54] N. Andersson and K. D. Kokkotas, *Mon. Not. R. Astron. Soc.* 299 (1998) 1059.

[55] S. Tsuruta, *Phys. Rep.* 292 (1998) 1.

[56] W. Becker ed., "*Neutron Stars and Pulsars*", Springer (2009).

[57] K. D. Kokkotas, T. A. Apostolatos, and N. Andersson, *Mon. Not. R. Astron. Soc.* 320 (2001) 307.

[58] H. Sotani, K. Tominaga, and K. I. Maeda, *Phys. Rev.* D65 (2001) 024010.

[59] H. Sotani and T. Harada, *Phys. Rev.* D68 (2003) 024019.

[60] H. Sotani, K. Kohri, and T. Harada, *Phys. Rev.* D69 (2004) 084008.

[61] E. Gaertig and K. D. Kokkotas, *Phys. Rev.* D80 (2009) 064026.

[62] B. Knippel, A. Sedrakian, *Phys. Rev.* D79 (2009) 083007.

[63] H. Sotani, N. Yasutake, T. Maruyama, and T. Tatsumi, *Phys. Rev.* D83 (2011) 024014.

[64] H. Sotani and K. D. Kokkotas, *Phys. Rev.* D70 (2004) 084026.

[65] H. Sotani and K. D. Kokkotas, *Phys. Rev.* D71 (2005) 124038.

[66] H. Sotani, *Phys. Rev.* D79 (2009) 064033.

[67] H. Sotani, *Phys. Rev.* D80 (2009) 064035.

[68] H. Sotani, *Phys. Rev.* D83 (2011) 124030.

[69] B. C. Barich, in "*Proceedings of the 17th International Conference on General Relativity and Gravitation*", edited by P. Florides, B. Nolan, and A. Ottewill (World Scientific, New Jersey, 2005).

[70] http://lisa.nasa.gov/.

[71] S. Kawamura *et al.*, “*Classical Quantum Gravity*” 23 (2006) S125,.

[72] A. L. Watts and T. E. Strohmayer, *Adv. Space Res.* 40 (2006) 1446.

[73] Y. Levin, *Mon. Not. R. Astron. Soc.* 368 (2006) L35.

[74] U. Lee, *Mon. Not. R. Astron. Soc.* 374 (2007) 1015.

[75] L. Samuelsson and N. Andersson, *Mon. Not. R. Astron. Soc.* 374 (2007) 256.

[76] H. Sotani, K. D. Kokkotas, and N. Stergioulas, *Mon. Not. R. Astron. Soc.* 375 (2007) 261.

[77] H. Sotani, K. D. Kokkotas, and N. Stergioulas, *Mon. Not. R. Astron. Soc.* 385 (2008) L5.

[78] H. Sotani, A. Colaiuda, and K. D. Kokkotas, *Mon. Not. R. Astron. Soc.* 385 (2008) 2161.

[79] H. Sotani and K. D. Kokkotas, *Mon. Not. R. Astron. Soc.* 395 (2009) 1163.

[80] M. Gabler *et al.*, *Mon. Not. R. Astron. Soc.* 410 (2011) L37.

[81] A. Colaiuda and K. D. Kokkotas, *Mon. Not. R. Astron. Soc.* 414 (2011) 3014.

[82] C.J. Pethick, A.Y. Potekhin *Phys. Lett.* B427 (1998) 7.

[83] A. W. Steiner and A. L. Watts, *Phys. Rev. Lett.* 103 (2009) 181101.

[84] H. Sotani, *Mon. Not. R. Astron. Soc.* 417 (2011) L70.

[85] M. Gearheart, W. G. Newton, J. Hooker, and B. A. Li, arXiv:1106.4875.

[86] H. Sotani, K. Nakazato, K. Iida, and K. Oyamatsu, *Phys. Rev. Lett.* 108 (2012) 201101.

[87] B. L. Schumaker and K. S. Thorne, *Mon. Not. R. Astron. Soc.* 203 (1983) 457.

[88] T. Strohmayer, H.M. van Horn, S. Ogata, H. Iyetomi, and S. Ichimaru, *Astrophys. J.* 375 (1991) 679 .

[89] N. K. Johnson-McDaniel, B. J. Owen, arXiv:1110.4650

[90] H. Sotani, T. Maruyama, and T. Tatsumi, arXiv:1207.4055.

[91] A. Burrows and J. M. Lattimer, *Astrophys. J.* 307 (1986) 178.

[92] T. Fischer *et al.* arXiv:1112.5528.

[93] B. L. Friman, and O. V. Maxwell, *Astrophys. J.*, 232 (1979) 541 .

[94] D. G. Yakovlev, O. Y. Gnedin, M. E. Gusakov, A. D. Kaminker, K. P. Levenfish, and A. Y. Potekhin, *Nucl. Phys.* A752 (2005) 590 .

[95] M. E. Gusakov, A. D. Kaminker, D. G. Yakovlev, and O. Y. Gnedin, *Mon. Not. R. Astron. Soc.* 363 (2005) 555 .

[96] J. M. Lattimer, C. J. Pethick, M. Prakash, and P. Haensel, *Phys. Rev. Lett.,* 66 (1991) 2701.

[97] D. G. Yakovlev, O. Y. Gnedin, A. D. Kaminker, K. P. Levenfish, A. Y. and Potekhin, *Adv. Space Res.* 33 (2004) 523.

[98] D. G. Yakovlev, A. D. Kaminker, and K. P. Levenfish, *Astron. Astrophys.*, 343 (1999) 650.

[99] M. E. Gusakov, *Astron. Astrophys.*, 389 (2002) 702.

[100] D. Page, J. M. Lattimer, M. Prakash, and A. W. Steiner, *Astrophys. J. Suppl.,* 115 (2004) 623.

[101] D. Page, J. M. Lattimer, M. Prakash, and A. W. Steiner, *Astrophys. J.*, 707, 1131 (2009).

[102] C. O. Heinke, P. G. Jonker, R. Wijnands, C. J. Deloye, and R. E. Taam, *Astrophys. J.*, 691 (2008) 7.

[103] W. C. G. Ho, and C. O. Heinke, *Nature*, 462 (2009) 71.

[104] C. O. Heinke, and W. C. G. Ho, *Astrophys. J.*, 719 (2010) L167.

[105] J. E. Reed, J. J. Hester, A. C. Fabian, and P. F. Winkler, *Astrophys. J.*, 440 (1995) 706.

[106] W. B. Ashworth, *J. Hist. Astron.*, 11 (1980) 1.

[107] D. G. Yakovlev, W. C. G. Ho, P. S. Shternin, C. O. Heinke, and A. Y. Potekhin, *Mon. Not. R. Astron. Soc.*, 411 (2011) 1977.

[108] D. Page, M. Prakash, J. M. Lattimer, and A. W. Steiner, *Phys. Rev. Lett.*, 106 (2011) 081101.

[109] P. S. Shternin, D. G. Yakovlev, C. O. Heinke, W. C. G. Ho, and D. J. Patnaude, *Mon. Not. R. Astron. Soc.*, 412 (2011) L108.

[110] N. Iwamoto, *Phys. Rev. Lett.*, 44 (1980) 1637.

[111] G. Baym, C. J. Pethick, and P. Sutherland, *Astrophys. J.*, 170 (1971) 299.

[112] D. Blaschke, H. Grigorian, and D. N. Voskresensky, *Astrophys. J.*, 533 (2000) 406.

[113] S. B. Rüster, V. Werth, M. Buballa, I. A. Shvkovy, and D. H. Rischke, *Phys. Rev.*, D72 (2005) 034004.

[114] A. Schmitt, "*Dense matter in compact stars*", Springer (2010).

[115] D. Hess, A. Sedrakian, *Phys. Rev.* D84 (2011) 063015.

[116] V. M. Kaspi, M. S. E. Roberts, and A. K. Harding, in "*Compact Stellar X-Ray Sources*", eds. W. H. G. Lewin and M. van der Klis (Cambridge University Press), (2006) 279-340.

[117] P. O. Slane, D. J. Helfand, and S. S. Murray, *Astrophys. J.*, 571 (2002) L45.

[118] M. C. Weisskopf, S. L. O'Dell, F. Paerels, R. F. Elsner, W. Kecker, A. F. Tennant, and D. A. Swartz, *Astrophys. J.*, 601 (2004) 1050.

[119] G. G. Pavlov, V. E. Zavlin, D. Sanwal, V. Burwitz, and G. P. Garmire, *Astrophys. J.*, 552 (2001) L129.

[120] W. F. Brisken, S. E. Thorsett, A. Golden, and W. M. Goss, *Astrophys. J.*, 593 (2003) L89.

[121] E. V. Gotthelf, J. P. Halpern, and R. Dodson, *Astrophys. J.*, 567 (2002) L125.

[122] B. M. Gaensler, N. S. Schulz, V. M. Kaspi, M. J. Pivovaroff, and W. E. Becker, *Astrophys. J.*, 588 (2003) 441.

[123] W. E. Becker, K. T. S. Brazier, and J. Truemper, *Astron. Astrophys.,* 306 (1996) 464.

[124] D. Page, http://www.astroscu.unam.mx/neutrones/NS-Data/data_good.dat (2002).

[125] T. Noda, M. Hashimoto, Y. Matsuo, N. Yasutake, T. Maruyama, T. Tatsumi, and M. Fujimoto, arXiv: 1109.1080.

In: Recent Advances in Quarks Research ISBN: 978-1-62257-970-9
Editors: H. Fujikage and K. Hyobanshi © 2013 Nova Science Publishers, Inc.

Chapter 5

QUARKS AND FRACTIONALLY QUANTIZED VORTICES IN SUPERCONDUCTORS - AN ANALOGY BETWEEN TWO WORLDS -

Takashi Yanagisawa
Electronics and Photonics Research Institute
National Institute of Advanced Industrial Science
and Technology (AIST) Tsukuba, Japan

Abstract

There are profound analogies between ideas in particle physics and the corresponding ones in condensed-matter physics. There is also an interesting analogy between quarks and fractionally quantized-flux vortices in superconductors. Half-quantized flux vortices exist in two-component superconductors under a magnetic field. They can be interpreted as a monopole, and two half-quantum flux vortices form a bound state connected by a domain wall. We can formulate vortices with 1/3-quantum flux in three-component superconductors with equivalent three bands. Under the duality transformation between charge and magnetic flux, vortices with 1/3-quantum magnetic flux can be regarded as quarks with the 1/3 charge. A three-vortex bound state is formulated and this bound state resembles a baryon in QCD. We have also a two-vortex bound state, formed by vortices with 1/3 and -1/3 quantized flux, respectively, that corresponds to a meson. Two or three fractionally quantized vortices are connected by domain walls and form a bound state. The quark confinement is discussed by analogy with baryonic bound states in superconductors.

1. Introduction

There is an interesting and profound analogy between particles physics and superconductivity. This was first pointed out by Youichiro Nambu, and he invented a concept of spontaneous symmetry breaking in particles physics. We say that spontaneous symmetry breaking occurred when the ground state of the system lost the invariance of the Hamiltonian of the system. There is no reason why an invariance of the Hamiltonian should be an invariance of the ground state. A good basic example is the Heisenberg model with spin-spin exchange interactions between nearest neighbors. The ground state is not rotationally invariant, with spins aligned in one direction, although the Hamiltonian is rotationally invariant. The ground state has a long-range order by breaking the rotational invariance. It is well known that the gapless Goldstone mode exists when the continuous symmetry is spontaneously broken. In the Heisenberg ferromagnet, this gapless mode is the spin wave excitation. There are many models that exhibit spontaneous symmetry breaking in the condensed-matter physics.

Superconductivity is most familiar phenomenon that occurs as a result of spontaneous symmetry breaking. This is described by the Ginzburg–Landau functional [1]:

$$F = \int d^3x \left[\alpha|\psi|^2 + \frac{1}{2}\beta|\psi|^4 + \frac{\hbar^2}{2m}\left|\left(\nabla + i\frac{2\pi}{\phi_0}\mathbf{A}\right)\psi\right|^2 + \frac{1}{8\pi}H^2 \right]. \tag{1.1}$$

ψ is the order parameter, $\beta > 0$ and α is written as $\alpha = \alpha_0(T - T_c)$. This free energy describes a spontaneous breaking of $U(1)$ symmetry. When $\alpha < 0$, that is $T < T_c$, the minimization of F yields a solution $\psi \neq 0$. The potential

$$V(\psi) = \alpha|\psi|^2 + \frac{1}{2}\beta|\psi|^4, \tag{1.2}$$

has an infinite number of possible minima for $\alpha < 0$ given by

$$\psi = \sqrt{-\frac{\alpha}{\beta}}\, e^{i\theta} \tag{1.3}$$

for any real angle θ in the range $0 \leq \theta \leq 2\pi$. Any choice of θ would have exactly the same energy that implies the existence of a massless Nambu–Goldstone boson.

There is a similarity between the Dirac equation and the gap equation of superconductivity. Nambu first noticed this fact [2]. The Dirac equation is written as

$$\left[p_0 - \begin{pmatrix} \mathbf{p}\cdot\sigma & 0 \\ 0 & -\mathbf{p}\cdot\sigma \end{pmatrix} - \begin{pmatrix} 0 & m \\ m & 0 \end{pmatrix}\right]\begin{pmatrix} \psi_R \\ \psi_L \end{pmatrix} = 0. \tag{1.4}$$

Here σ_j $(j = 1,2,3)$ are Pauli matrices. If the mass m vanishes, the Dirac equation possesses chiral symmetry and is invariant under the transformation

$$\psi \to e^{i\alpha\gamma_5}\psi, \tag{1.5}$$

for $\psi = (\psi_R, \psi_L)^t$ and

$$\gamma_5 = \begin{pmatrix} I & 0 \\ 0 & -I \end{pmatrix}. \tag{1.6}$$

The mass generation occurs as a result of spontaneous breaking of chiral symmetry [3,4]. The mass is determined by the self-consistency equation (Figure 1). This equation has a close analogy with superconductivity; the gap is generated by spontaneous breaking of $U(1)$ symmetry [5,6]. The equation for paired electrons is written as

$$\left[p_0 - \begin{pmatrix} \xi_p & 0 \\ 0 & -\xi_p \end{pmatrix} - \begin{pmatrix} 0 & \Delta \\ \Delta & 0 \end{pmatrix}\right]\begin{pmatrix} \psi_{p\uparrow} \\ \psi^\dagger_{-p\downarrow} \end{pmatrix} = 0, \tag{1.7}$$

where ξ_p is the kinetic energy measured from the Fermi surface and Δ is the energy gap. This equation gives the eigenvalues

$$p_0 = E_p = \pm\sqrt{\xi_p^2 + \Delta^2}. \tag{1.8}$$

Therefore there is a close analogy between the Dirac equation in elementary particle theory and the gap equation in the theory of superconductivity.

We show a list of correspondences between superconductivity and particle physics in Table 1. The gapless mode emerging from the spontaneously broken global U(1) symmetry in superconductors is known as the Anderson–Bogoliubov mode. TRS breaking denotes time-reversal symmetry breaking which occurs in a triplet-pairing superconductor or in a three-band superconductor.

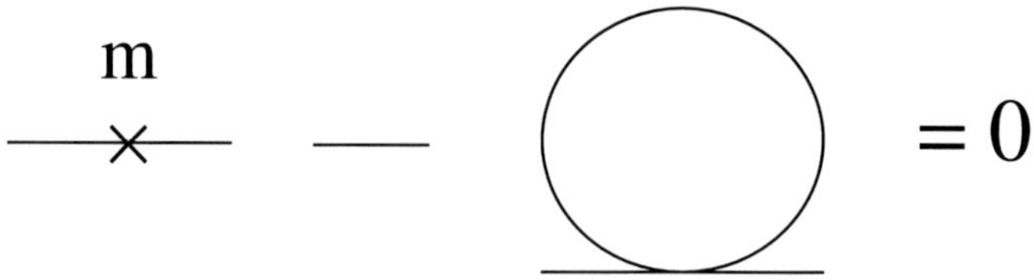

Figure 1. Self-consistency equation for the mass.

2. Mass and Gap Equations

We discuss a similarity between superconductivity and particle physics in the gap equation.

2.1. Superconductivity

Let us consider the BCS Hamiltonian:

$$H = \int d\mathbf{r} \sum_{\sigma} \psi_{\sigma}^{\dagger}(\mathbf{r}) \left(\frac{\mathbf{p}^2}{2m} - \mu \right) \psi_{\sigma}(\mathbf{r}) - g \int d\mathbf{r} \psi_{\uparrow}^{\dagger}(\mathbf{r}) \psi_{\downarrow}^{\dagger}(\mathbf{r}) \psi_{\downarrow}(\mathbf{r}) \psi_{\uparrow}(\mathbf{r}), \quad (2.1)$$

where σ is the spin index $\uparrow$ and $\downarrow$, μ is the chemical potential and $g > 0$ is the coupling constant of the attractive interaction. In the momentum space, this is written as

$$H = \sum_{k\sigma} \xi_k c_{k\sigma}^{\dagger} c_{k\sigma} - g \frac{1}{V} \sum_{kk'q} c_{k'\uparrow}^{\dagger} c_{-k'+q\downarrow}^{\dagger} c_{-k+q\downarrow} c_{k\uparrow}, \quad (2.2)$$

where $\xi_k = \varepsilon_k - \mu$ for the electron dispersion ε_k. The corresponding Lagrangian density is

$$\mathcal{L} = \sum_{\sigma} \psi_{\sigma}^{\dagger}(x) \left(i\hbar \frac{\partial}{\partial t} + \frac{\hbar^2}{2m} \nabla^2 + \mu \right) \psi_{\sigma}(x) + g \psi_{\uparrow}^{\dagger}(x) \psi_{\downarrow}^{\dagger}(x) \psi_{\downarrow}(x) \psi_{\uparrow}(x). \quad (2.3)$$

Using the Nambu notation [7],

$$\psi = \begin{pmatrix} \psi_{\uparrow} \\ \psi_{\downarrow}^{\dagger} \end{pmatrix}, \quad (2.4)$$

Table 1.

Superconductors	Particles
Broken symmetry	Broken symmetry
Ground state: BCS	Vacuum
Energy gap	Mass
Quasi particles	Particles
Gapless mode (Anderson–Bogoliubov)	Nambu–Goldstone mode
Plasma mode	Higgs mechanism
Meissner effect	Higgs mechanism
TRS breaking	CP violation
Half-quantum vortex	Monopole
Fractional vortices	Quarks
Leggett mode	?

the Lagrangian density becomes

$$\mathcal{L} = \psi^\dagger \left(\sigma_0 i\hbar \frac{\partial}{\partial t} - \sigma_3 \xi(\nabla) \right) \psi - \frac{g}{4} \left[(\psi^\dagger \psi)^2 - (\psi^\dagger \sigma_3 \psi)^2 \right], \tag{2.5}$$

where σ_0 is the unit matrix and $\xi(\nabla) = -\hbar^2 \nabla^2/(2m) - \mu = \mathbf{p}^2/(2m) - \mu$. The vacuum partition function is represented by a functional integral,

$$Z = \int d\psi^\dagger d\psi \exp\left(\frac{i}{\hbar} \int d^d x \mathcal{L} \right). \tag{2.6}$$

d is the space-time dimension. This can be written in a bilinear form by applying a Hubbard–Stratonovich transformation,

$$\begin{aligned} &\exp\left(\frac{i}{\hbar} g \int d^d x \psi_\uparrow^\dagger \psi_\downarrow^\dagger \psi_\downarrow \psi_\uparrow \right) \\ &= \int d\Delta^* d\Delta \exp\left[-\frac{i}{\hbar} \int d^d x \left(\Delta^* \psi_\downarrow \psi_\uparrow + \Delta \psi_\uparrow^\dagger \psi_\downarrow^\dagger + \frac{1}{g} |\Delta|^2 \right) \right], \end{aligned} \tag{2.7}$$

where Δ^* and Δ are auxiliary fields and an overall normalization factor is excluded. The partition function has the form

$$Z = \int d\psi^\dagger d\psi \int d\Delta^* d\Delta \exp\left(\frac{i}{\hbar}\int d^d x \mathcal{L}_{eff}\right), \tag{2.8}$$

where

$$\mathcal{L}_{eff} = \psi^\dagger \left[\sigma_0 i\hbar\frac{\partial}{\partial t} - \sigma_3\xi(\nabla) - \begin{pmatrix} 0 & \Delta \\ \Delta^* & 0 \end{pmatrix}\right]\psi - \frac{1}{g}|\Delta|^2. \tag{2.9}$$

The field equations obtained by variation of the Lagrangian are

$$\left[i\hbar\frac{\partial}{\partial t} - \sigma_3\xi(\nabla) - \begin{pmatrix} 0 & \Delta \\ \Delta^* & 0 \end{pmatrix}\right]\psi = 0, \tag{2.10}$$

$$\Delta = g\psi_\uparrow\psi_\downarrow. \tag{2.11}$$

The equation for Δ shows that Δ describes a pair of electrons that forms a spin-singlet. If we approximate Δ by its average $\bar{\Delta} = g\langle\psi_\uparrow\psi_\downarrow\rangle$, we obtain a self-consistency equation for $\bar{\Delta}$. By performing the Grassman integration over the fields $\psi^\dagger$ and ψ, we obtain the effective action

$$S(\Delta^*,\Delta) = -\frac{1}{g}\int d^d x|\Delta(x)|^2 - i\hbar\mathrm{Tr}\ln\begin{pmatrix} p_0-\xi(\mathbf{p}) & -\Delta(x) \\ -\Delta^*(x) & p_0+\xi(\mathbf{p}) \end{pmatrix}, \tag{2.12}$$

for which the partition function is

$$Z = \int d\Delta^* d\Delta \exp\left(\frac{i}{\hbar}S(\Delta^*,\Delta)\right). \tag{2.13}$$

Now the averaged value $\bar{\Delta}$ of the gap function Δ is determined by adopting the saddle point approximation. The field equation reads

$$\frac{\delta S(\bar{\Delta}^*,\bar{\Delta})}{\delta\bar{\Delta}^*} = 0. \tag{2.14}$$

We obtain a solution assuming that $\bar{\Delta} > 0$ is a constant. This yields

$$\frac{1}{g}\bar{\Delta} = i\hbar\mathrm{Tr}G_0(p)\begin{pmatrix} 0 & 0 \\ 1 & 0 \end{pmatrix}, \tag{2.15}$$

where $G_0(p)$ is the Green function including $\bar{\Delta}$,

$$\begin{aligned} G_0(p) &= \begin{pmatrix} p_0-\xi(\mathbf{p}) & -\bar{\Delta} \\ -\bar{\Delta}^* & p_0+\xi(\mathbf{p}) \end{pmatrix}^{-1} \\ &= \frac{1}{p_0^2 - E(\mathbf{p})^2 + i\delta}\begin{pmatrix} p_0+\xi(\mathbf{p}) & \bar{\Delta} \\ \bar{\Delta}^* & p_0-\xi(\mathbf{p}) \end{pmatrix}. \end{aligned} \tag{2.16}$$

Here,

$$E(\mathbf{p}) = \sqrt{\xi(\mathbf{p})^2 + \bar{\Delta}^2} \tag{2.17}$$

is the single-particle excitation energy. Then we obtain the gap equation

$$\frac{1}{g} = \frac{1}{2}\int \frac{d^3k}{(2\pi)^3}\frac{1}{E(\mathbf{k})}. \tag{2.18}$$

The superconducting gap is

$$\bar{\Delta} = 2\hbar\omega_D \exp\left(-\frac{1}{\rho g}\right), \tag{2.19}$$

with the energy cutoff $\hbar\omega_D$ and the density of states ρ at the Fermi energy.

2.2. Nambu–Jona–Lasinio Model

The Nambu–Jona–Lasinio model is

$$\mathcal{L} = \bar{\psi} i\gamma^\mu \partial_\mu \psi + g\left[(\bar{\psi}\psi)^2 - (\bar{\psi}\gamma_5\psi)^2\right], \tag{2.20}$$

which has the form similar to the BCS model. We set $\hbar = 1$ in this section. γ_μ and γ_5 are Dirac gamma matrices. This Lagrangian is invariant under the particle number and chiral transformations,

$$\psi \to \exp(i\alpha)\psi, \quad \bar{\psi} \to \bar{\psi}\exp(-i\alpha), \tag{2.21}$$

$$\psi \to \exp(i\gamma_5\alpha)\psi, \quad \bar{\psi} \to \bar{\psi}\exp(i\gamma_5\alpha). \tag{2.22}$$

In a similar way after the spontaneous symmetry breaking, the fermion (nucleon) acquires a mass $m \propto 2g\langle\bar{\psi}\psi\rangle$.

Using an identity

$$1 = \text{const.}\int d\sigma' d\pi' \exp i\int d^4x \left[-\frac{1}{4g}(\sigma'^2 + \pi'^2)\right], \tag{2.23}$$

the partition function is written as

$$\begin{aligned} Z &= \int d\bar{\psi} d\psi d\sigma' d\pi' \\ &\times \exp i\int d^4x \left[\bar{\psi} i\gamma^\mu \partial_\mu \psi + g\left((\bar{\psi}\psi)^2 - (\bar{\psi}\gamma_5\psi)^2\right) - \frac{1}{4g}(\sigma'^2 + \pi'^2)\right]. \end{aligned} \tag{2.24}$$

We define new σ and π fields by

$$\begin{aligned} \sigma' &= \sigma + 2g\bar{\psi}\psi, && (2.25)\\ \pi' &= \pi + 2gi\bar{\psi}\gamma_5\psi, && (2.26) \end{aligned}$$

then we have

$$Z = \int \bar{\psi} d\psi d\sigma d\pi \exp\left(i\int d^4x \mathcal{L}_{eff}\right), \tag{2.27}$$

where

$$\begin{aligned} \mathcal{L}_{eff} &= \mathcal{L} - \frac{1}{4g}\left[(\sigma + 2g\bar{\psi}\psi)^2 + (\pi + 2gi\bar{\psi}\gamma_5\psi)^2\right] \\ &= \bar{\psi}\left[i\gamma^\mu\partial_\mu - (\sigma + i\gamma_5\pi)\right]\psi - \frac{1}{4g}(\sigma^2 + \pi^2). \end{aligned} \tag{2.28}$$

Then we obtain the effective action

$$\begin{aligned} S_{eff}(\sigma,\pi) &= -i\ln\det\left[i\gamma^\mu\partial_\mu - (\sigma + i\gamma_5\pi)\right] - \frac{1}{4g}\int d^4x(\sigma^2 + \pi^2) \\ &= -i\,\mathrm{Tr}\ln\left[i\gamma^\mu\partial_\mu - (\sigma + i\gamma_5\pi)\right] - \frac{1}{4g}\int d^4x(\sigma^2 + \pi^2). \end{aligned} \tag{2.29}$$

The saddle point approximation leads to a solution such that $\sigma = \sigma_0$ is a constant and $\pi = 0$. The equation for σ_0 is

$$\sigma_0 = 2ig\,\mathrm{Tr}\frac{1}{i\gamma^\mu\partial_\mu - \sigma_0}. \tag{2.30}$$

Because σ_0 is the mass m of the fermion ψ, the mass m is determined by

$$1 = 8gi\int \frac{d^4k}{(2\pi)^4}\frac{1}{k^2 - m^2} = 8g\int \frac{d^4k_E}{(2\pi)^4}\frac{1}{k_E^2 + m^2}, \tag{2.31}$$

which has a nontrivial solution $m \neq 0$ when

$$0 < \frac{2\pi^2}{g\Lambda^2} < 1. \tag{2.32}$$

In this case the mass m is determined by

$$\frac{2\pi^2}{g\Lambda^2} = 1 - \frac{m^2}{\Lambda^2}\ln\left(1 + \frac{\Lambda^2}{m^2}\right). \tag{2.33}$$

We define the field h_σ by

$$\sigma = \sigma_0 + h_\sigma, \tag{2.34}$$

then h_σ is a massive boson whose mass is $2m$. The field π represents a massless boson which is the Nambu–Goldstone boson. The result that the field h_σ acquires the mass $2m$ is well understood by an analogy to superconductivity, where the excitation energy is $2\bar{\Delta}$ when a pair of electrons is excited above the Fermi energy. The model has been generalized to a more realistic two-flavor model:

$$\mathcal{L} = \bar{\psi} i\gamma^\mu \partial_\mu \psi + g\left[(\bar{\psi}\psi)^2 - \sum_i (\bar{\psi}\gamma_5\tau_i\psi)(\bar{\psi}\gamma_5\tau_i\psi)\right]. \tag{2.35}$$

As is obvious from the discussion here, the nucleon mass generation is very analogous to the gap generation in superconductors. We note that the mass is finite only when $0 < 2\pi^2/g\Lambda^2 < 1$ holds in the Nambu–Jona–Lasinio model while the superconducting gap always exists as far as $g > 0$.

3. Multi-band Superconductors

We consider the multi-band BCS model with the attractive interactions:

$$\begin{aligned} H &= \sum_{i\sigma} \int d\mathbf{r} \psi_{i\sigma}^\dagger(\mathbf{r}) K_i(\mathbf{r}) \psi_{i\sigma}(\mathbf{r}) \\ &- \sum_{ij} g_{ij} \int d\mathbf{r} \psi_{i\uparrow}^\dagger(\mathbf{r}) \psi_{i\downarrow}^\dagger(\mathbf{r}) \psi_{j\downarrow}(\mathbf{r}) \psi_{j\uparrow}(\mathbf{r}), \end{aligned} \tag{3.1}$$

where i and j $(= 1, 2, \ldots)$ are band indices. $K_i(\mathbf{r})$ stands for the kinetic operator. We assume that $g_{ij} = g_{ji}^*$. The second term is the pairing interaction and g_{ij} are coupling constants. The mean-field Hamiltonian is

$$\begin{aligned} H_{MF} = \sum_i \int d\mathbf{r} \Bigg[& \sum_\sigma \psi_{i\sigma}^\dagger(\mathbf{r}) K_i(\mathbf{r}) \psi_{i\sigma}(\mathbf{r}) \\ & + \Delta_i(\mathbf{r}) \psi_{i\uparrow}^\dagger(\mathbf{r}) \psi_{i\downarrow}^\dagger(\mathbf{r}) + \Delta_i^*(\mathbf{r}) \psi_{i\downarrow}(\mathbf{r}) \psi_{i\uparrow}(\mathbf{r}) \Bigg], \end{aligned} \tag{3.2}$$

where the gap function in each band is defined by

$$\Delta_i(\mathbf{r}) = -\sum_j g_{ij} \left\langle \psi_{j\downarrow}(\mathbf{r}) \psi_{j\uparrow}(\mathbf{r}) \right\rangle, \tag{3.3}$$

and its complex conjugate is

$$\Delta_i^*(\mathbf{r}) = -\sum_j g_{ji} \left\langle \psi_{j\uparrow}^\dagger(\mathbf{r})\psi_{j\downarrow}^\dagger(\mathbf{r}) \right\rangle. \tag{3.4}$$

We define Green's functions as follows [8],

$$G_{j\sigma\sigma'}(x-x') = -\left\langle T_\tau \psi_{j\sigma}(x)\psi_{j\sigma'}^\dagger(x') \right\rangle, \tag{3.5}$$

$$F_{j\sigma\sigma'}^+(x-x') = \left\langle T_\tau \psi_{j\sigma}^\dagger(x)\psi_{j\sigma'}^\dagger(x') \right\rangle, \tag{3.6}$$

where T_τ is the time-ordering operator and we use the notation $x = (\tau, \mathbf{r})$. In terms of the Green's functions, the gap functions satisfy the system of equations

$$\Delta_i^*(\mathbf{r}) = \sum_j g_{ij}^* F_{j\downarrow\uparrow}^+(\tau' = \tau + 0; \mathbf{r}, \mathbf{r}) = \sum_j g_{ij}^* \frac{1}{\beta} \sum_n F_{j\downarrow\uparrow}^+(i\omega_n; \mathbf{r}, \mathbf{r}). \tag{3.7}$$

This yields the gap equation,

$$\Delta_i = \sum_j g_{ij} N_j \Delta_j \int d\xi_j \frac{1}{E_j} \tanh\left(\frac{E_j}{2T}\right), \tag{3.8}$$

where $E_j = \sqrt{\xi_j^2 + |\Delta_j|^2}$ and T is the temperature where we set Boltzmann constant k_B to unity. N_j is the density of states at the Fermi surface. Since all the bands couple with each other through mutual interactions g_{ij}, we have one critical temperature T_c [9]. At the critical temperature $T = T_c$, this equation reads [10]1

$$\Delta_i = \ln\left(\frac{2e^\gamma \omega_c}{\pi T_c}\right) \sum_j g_{ij} N_j \Delta_j, \tag{3.9}$$

for the cutoff energy ω_c. γ denotes the Euler constant. Here we assume the same cutoff energy in all the interactions. The system of equations in Eq. (3.7) yields a set of differential equations [10]

$$\begin{aligned} \Delta_j^*(\mathbf{r}) &= \ln\left(\frac{2e^\gamma \omega_c}{\pi T}\right) \sum_\ell g_{j\ell}^* N_\ell \Delta_\ell^*(\mathbf{r}) \\ &+ \frac{7\zeta(3)}{48(\pi T_c)^2} \sum_\ell g_{j\ell}^* N_\ell v_\ell^2 \left(\nabla + i\frac{2e}{\hbar c}\mathbf{A}\right)^2 \Delta_\ell^*(\mathbf{r}) \\ &- \frac{7\zeta(3)}{8(\pi T_c)^2} \sum_\ell g_{j\ell}^* N_\ell \Delta_\ell^*(\mathbf{r}) |\Delta_\ell(\mathbf{r})|^2. \end{aligned} \tag{3.10}$$

Here, e is the charge of the electron, and v_ℓ is the electron velocity at the Fermi surface in the ℓ-th band. The Planck constant $\hbar$ has been dropped except in front of the vector potential $\mathbf{A}$. We set $a_{mn} = g_{mn}N_n$. Then, the equation for Δ_j is

$$0 = \left[a_{jj}\ln\left(\frac{2e^\gamma\omega_c}{\pi T}\right) - 1\right]\Delta_j + \ln\left(\frac{2e^\gamma\omega_c}{\pi T}\right)\sum_{\ell(\neq j)} a_{j\ell}\Delta_\ell + \cdots. \quad (3.11)$$

In the second term of the right-hand side we can replace T by T_c near the transition temperature. We define the matrix $A = (a_{ij})$ and its inverse $A^{-1} = (b_{j\ell})$. Then the equation for Δ_j reads

$$0 = \left[a_{jj}\ln\left(\frac{2e^\gamma\omega_c}{\pi T}\right) - 1\right]\Delta_j + \sum_{\ell(\neq j)} a_{j\ell}\sum_m b_{\ell m}\Delta_m + \cdots. \quad (3.12)$$

For example, for $j = 1$ we obtain

$$0 = g_{11}\left[\left(N_1\ln\left(\frac{2e^\gamma\omega_c}{\pi T}\right) - \frac{1}{\det G}(G^{-1})_{11}\right)\Delta_1 - \frac{1}{\det G}(G^{-1})_{12}\Delta_2 - \frac{1}{\det G}(G^{-1})_{13}\Delta_3\right] + \cdots, \quad (3.13)$$

where $G = (g_{ij})$ is the matrix of coupling constants. To obtain the multi-band Ginzburg–Landau functional, we multiply Eq. (3.12) by $\Delta_j^* N_j$ and take a summation with respect to j. We use the gap equation $\Delta_\ell^* = \eta\sum_j g_{\ell j}^* N_j\Delta_j^*$ at $T = T_c$. Then the energy functional density f is

$$\begin{aligned} f = & - \sum_j\left(N_j\ln\frac{2e^\gamma\omega_c}{\pi T} - (G^{-1})_{jj}\right)|\Delta_j|^2 + \sum_{j\ell}\Delta_j^*(G^{-1})_{j\ell}\Delta_\ell \\ & - \frac{7\zeta(3)}{48\pi^2T_c^2}\sum_\ell N_\ell v_\ell^2\Delta_\ell^*\left(\nabla - i\frac{2e}{\hbar c}\mathbf{A}\right)^2\Delta_\ell \\ & + \frac{7\zeta(3)}{16\pi^2T_c^2}\sum_\ell N_\ell|\Delta_\ell|^4. \end{aligned} \quad (3.14)$$

Here we neglected unimportant constants. The fourth order term is simply given by $(|\Delta_\ell|^2)^2$. We can also derive this functional using the functional integral method.

4. Abelian Projection and a Singularity [11]

Let us consider a field $X(x)$. We assume that $X(x)$ is hermitian and is transformed as follows.

$$X(x) \to U(x)X(x)U(x)^{\dagger}. \tag{4.1}$$

We adopt that $X(x)$ is diagonalized:

$$X(x) = \begin{pmatrix} \lambda_1(x) & \cdots & \\ \vdots & \ddots & \vdots \\ & \cdots & \lambda_N(x) \end{pmatrix}, \tag{4.2}$$

where $\lambda_j(x)$ $(j = 1, \ldots, N)$ are eigenvalues of $X(x)$. If all the $\lambda_j(x)$ are different each other, $X(x)$ is invariant under the transformation $X(x) \to U(x)X(x)U(x)^{\dagger}$, where

$$U(x) = \begin{pmatrix} e^{i\theta_1(x)} & \cdots & \\ \vdots & \ddots & \vdots \\ & \cdots & e^{i\theta_N(x)} \end{pmatrix}. \tag{4.3}$$

Here $\sum_{i=1}^{N} \theta_i(x) = 0$ and we assume that $[X(x), U(x)] = 0$. $U(x)$ is an element of $U(1)^{N-1}$.

Let us assume that two of $\lambda_j(x)$ are degenerate, for example, $\lambda_i(x) = \lambda_j(x)$ for $i \neq j$. Then we have $SU(2)$ symmetry at this point. It is known that a monopole-like singularity appears at the point where the eigenvalues are degenerate. The same singularity appears in a vortex state with fractionally quantized flux in a multi-band superconductor. Let us consider the $SU(2)$ case. The matrix $U(x)$ is written as, using Pauli matrices σ_j $(j = 1, 2, 3)$,

$$\begin{aligned} U(x) &= e^{i\xi(x)\sigma_3/2} e^{i\theta(x)\sigma_2/2} e^{i\varphi(x)\sigma_3/2} \\ &= \begin{pmatrix} e^{i(\varphi+\xi)(x)/2} \cos\dfrac{\theta(x)}{2} & e^{-i(\varphi-\xi)(x)/2} \sin\dfrac{\theta(x)}{2} \\ -e^{i(\varphi-\xi)(x)/2} \sin\dfrac{\theta(x)}{2} & e^{-i(\varphi+\xi)(x)/2} \cos\dfrac{\theta(x)}{2} \end{pmatrix}, \end{aligned} \tag{4.4}$$

where ξ, θ and φ are angles in the range of $0 \leq \xi \leq 2\pi$, $0 \leq \theta \leq \pi$ and $0 \leq \varphi \leq 2\pi$.

θ and φ are Euler angles. $U^\dagger(x)$ is given by

$$U^\dagger(x) = \begin{pmatrix} e^{-i(\varphi+\xi)(x)/2}\cos\frac{\theta(x)}{2} & -e^{-i(\varphi-\xi)(x)/2}\sin\frac{\theta(x)}{2} \\ e^{i(\varphi-\xi)(x)/2}\sin\frac{\theta(x)}{2} & e^{i(\varphi+\xi)(x)/2}\cos\frac{\theta(x)}{2} \end{pmatrix}. \tag{4.5}$$

We define

$$\Omega_\mu(x) = i\frac{1}{g}U(x)\partial_\mu U^\dagger(x). \tag{4.6}$$

Then, under the gauge transformation, the gauge field $A_\mu(x)$ is transformed to

$$A_\mu(x) \to A'_\mu(x) = U(x)A_\mu(x)U^\dagger(x) + \Omega_\mu(x). \tag{4.7}$$

Ω_μ is evaluated as

$$\begin{aligned} \Omega_\mu &= \frac{1}{g}\left[\frac{1}{2}(\partial_\mu\varphi\cos\theta)\sigma_3 + \left(-\frac{1}{2}\partial_\mu\varphi\cos\xi\sin\theta + \frac{1}{2}\sin\xi\partial_\mu\theta\right)\sigma_1\right. \\ &+ \left.\left(\frac{1}{2}\partial_\mu\varphi\sin\xi\sin\theta + \frac{1}{2}\cos\xi\partial_\mu\theta\right)\sigma_2\right]. \end{aligned} \tag{4.8}$$

When $\xi = -\varphi$, the diagonal term, being proportional to σ_3, is

$$\frac{1}{2g}\partial_\mu\varphi(\cos\theta - 1) = -\frac{1}{2g}\left(-\frac{y}{r(r+z)}, \frac{x}{r(r+z)}, 0\right), \tag{4.9}$$

for $r = \sqrt{x^2+y^2+z^2}$. This term has a singularity $z = -r$ ($\theta = \pi$) and represents a Dirac monopole. The corresponding gauge field is

$$\mathbf{A}_I = \frac{q_m}{4\pi}\left(-\frac{y}{r(r+z)}, \frac{x}{r(r+z)}, 0\right) = \frac{q_m}{4\pi}\frac{1-\cos\theta}{r\sin\theta}\mathbf{e}_\varphi, \tag{4.10}$$

where q_m is the magnetic charge and $\mathbf{e}_\varphi = (-\sin\varphi, \cos\varphi, 0)$. The magnetic field is

$$\mathbf{B}_m = \nabla\times\mathbf{A}_I = \frac{q_m}{4\pi r^2}\mathbf{e}_r, \tag{4.11}$$

where $\mathbf{e}_r = \mathbf{r}/r = (x,y,z)/r$. The gauge field which has a singularity at $z=r$ is

$$\mathbf{A}_{II} = \frac{q_m}{4\pi}\left(\frac{y}{r(r-z)}, -\frac{x}{r(r-z)}, 0\right) = -\frac{q_m}{4\pi}\frac{1+\cos\theta}{r\sin\theta}\mathbf{e}_\varphi. \tag{4.12}$$

This satisfies

$$\nabla \times \mathbf{A}_{II} = \frac{q_m}{4\pi r^2}\mathbf{e}_r. \tag{4.13}$$

$\mathbf{A}_I$ and $\mathbf{A}_{II}$ are connected by the gauge transformation,

$$\mathbf{A}_I = \mathbf{A}_{II} + \nabla\left(\frac{q_m}{2\pi}\varphi\right). \tag{4.14}$$

We will see that the gauge field for monopole indeed appears in a multi-band superconductor.

5. Half-Quantum Vortex

In this section we consider a half-quantum flux vortex in a two-band superconductor [12–14] and discuss an analogy with a monopole. We write the order parameters as

$$\psi_j = \rho_j e^{i\theta_j}, \tag{5.1}$$

where $\rho_j = |\psi_j|$ is a real quantity. For simplicity, we assume that the coefficients of the Josephson terms are real: $\gamma_{ij} = \gamma_{ji}^* = \gamma_{ji}$. The free energy density is denoted as f, that is, the free energy is given by the integral of f over the space. f is written as

$$\begin{aligned} f &= \sum_j \alpha_j\rho_j^2 + \frac{1}{2}\sum_j \beta_j\rho_j^4 - 2\gamma_{12}\rho_1\rho_2\cos(\theta_1-\theta_2) \\ &- \sum_j K_j\rho_j e^{-i\theta_j}\left(\nabla + i\frac{2\pi}{\phi_0}\mathbf{A}\right)^2(\rho_j e^{i\theta_j}) + \frac{1}{8\pi}\mathbf{H}^2. \end{aligned} \tag{5.2}$$

We focus on the role of phases of the order parameters; we assume that

$$\rho_j = \rho, \;\; K_j = K, \tag{5.3}$$

and define new phase variables

$$\Phi = \theta_1 + \theta_2, \;\; \phi = \theta_1 - \theta_2. \tag{5.4}$$

The free energy density is

$$f = 2\alpha\rho^2 + \beta\rho^4 - 2\rho^2\gamma_{12}\cos(\phi) - 2K\rho\nabla^2\rho + 2K\frac{4\pi^2}{\phi_0^2}\rho^2\mathbf{A}^2$$

$$+ \quad \frac{1}{2}K\rho^2(\nabla\Phi)^2 + K\frac{4\pi}{\phi_0}\rho^2\mathbf{A}\cdot\nabla\Phi + \frac{1}{2}K\rho^2(\nabla\phi)^2 + \frac{1}{8\pi}\mathbf{H}^2, \tag{5.5}$$

where $\alpha = (1/2)\sum_j \alpha_j$ and $\beta = (1/2)\sum_j \beta_j$.

Let us focus on the phase difference ϕ. Excitations that are connected to this variable is sometimes called Leggett mode [15]. The equation of motion for ϕ reads

$$2\gamma_{12}\rho^2 \sin\phi - K\rho^2\nabla^2\phi = 0. \tag{5.6}$$

We define $\alpha = 2\gamma_{12}/K$ and adopt that α is positive. The sign of α does not matter because we can change the sign of $\sin\phi$ by shifting the variable ϕ. We consider a one-dimensional-like solution where ϕ has spatial dependence only in one direction, for example, in x direction. In this case the equation for ϕ is

$$\frac{d^2\phi}{dx^2} = \alpha\sin\phi. \tag{5.7}$$

We use the boundary condition such that $\phi \to 0$ as $x \to -\infty$ and $\phi \to 2\pi$ as $x \to \infty$. Then we have a kink solution:

$$\phi = \pi + 2\sin^{-1}\left(\tanh(\sqrt{\alpha}x)\right). \tag{5.8}$$

The phase difference ϕ changes from 0 to 2π across the kink. This means that θ_1 changes from 0 to π and at the same time θ_2 changes from 0 to $-\pi$. In this case, a half-quantum-flux vortex exists at the end of the kink. This is shown in Figure 2 where the half-quantum vortex is at the edge of the cut (kink). A net change of θ_1 is 2π by a counterclockwise encirclement of the vortex, and that of θ_2 vanishes. Then, we have a half-quantum flux vortex.

The half-quantum vortex can be interpreted as a monopole. Let us assume that there is a cut, namely, kink on the real axis for $x > 0$. The phase θ_1 is represented by

$$\theta_1 = \frac{1}{2}\mathrm{Im}\log\zeta, \tag{5.9}$$

where

$$\zeta = x + iy. \tag{5.10}$$

The singularity of θ_j can be transferred to a singularity of the gauge field by a gauge transformation. Let us consider the fictitious z axis perpendicular to the $x-y$ plane. The gauge potential (1-form) is given by

$$A_{\pm} = -\frac{1}{2}\frac{1}{r(z\pm r)}(ydx - xdy) = \frac{1}{2}(\pm 1 - \cos\theta)d\varphi, \tag{5.11}$$

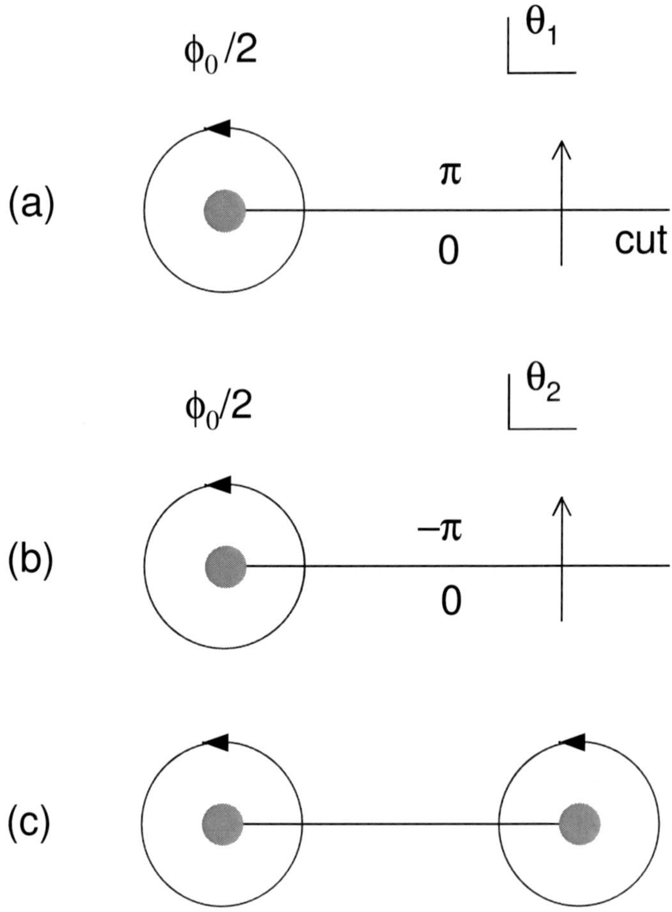

Figure 2. Half-quantum flux vortex with line singularity. The phase variables θ_1 and θ_2 have line singularities, as shown in (a) and (b). A pair of two half-flux vortices connected by the singularity (domain wall) is shown in (c).

where $r = \sqrt{x^2+y^2+z^2}$, and θ and φ are Euler angles. $A_\pm$ correspond to the gauge potential in the upper and lower hemisphere $H_\pm$, respectively. $A_\pm$ are connected by

$$A_+ = A_- + d\varphi. \tag{5.12}$$

This is the U(1) bundle P over the sphere S^2. At $z = 0$ A_+ coincides with the gauge field A for half-quantum vortex. The Chern class is defined as

$$c_1(P) = -\frac{1}{2}F = -\frac{1}{2}dA_+. \tag{5.13}$$

The Chern number is given as

$$C_1 = \int_{S^2} c_1 = -\frac{1}{2\pi}\int_{S^2} F = -\frac{1}{2\pi}\left[\int_{H_+} dA_+ \int_{H_-} dA_-\right] = -1. \tag{5.14}$$

6. Time-Reversal Symmetry Breaking

Let us consider the case $N = 3$. When two eigenvalues are degenerate, we have a singularity. This occurs in a three-band superconductor and leads to time-reversal symmetry breaking. This may be seen as an analogy to the Kobayashi–Maskawa matrix [16]. To see this, we set

$$X = \begin{pmatrix} g_{11}N_1 & g_{12}N_2 & g_{13}N_3 \\ g_{21}N_1 & g_{22}N_2 & g_{23}N_3 \\ g_{31}N_1 & g_{32}N_2 & g_{33}N_3 \end{pmatrix}. \tag{6.1}$$

Then the gap equation is

$$\lambda \begin{pmatrix} \Delta_1 \\ \Delta_2 \\ \Delta_3 \end{pmatrix} = X \begin{pmatrix} \Delta_1 \\ \Delta_2 \\ \Delta_3 \end{pmatrix}, \tag{6.2}$$

where $1/\lambda = \ln(2e^{\gamma}\omega_c/\pi T_c)$. We assume that the matrix $(g_{ij}N_j)$ is real and symmetric. When two eigenvalues are degenerate, we obtain two complex eigenvectors by making linear combinations. For example, let us consider a matrix $X = (X_{ij})$ given by

$$X_{11} = J(\cos^2\theta\cos^2\varphi + \sin^2\varphi) + D, \tag{6.3}$$

$$X_{12} = X_{21} = -J\sin^2\theta\sin\varphi\cos\varphi, \tag{6.4}$$

$$X_{13} = X_{31} = -J\sin\theta\cos\theta\cos\varphi, \tag{6.5}$$

$$X_{22} = J(\cos^2\theta\sin^2\varphi + \cos^2\varphi) + D, \tag{6.6}$$

$$X_{23} = X_{32} = -J\sin\theta\cos\theta\sin\varphi, \tag{6.7}$$

$$X_{33} = J\sin^2\theta + D, \tag{6.8}$$

where J is a positive constant and D is also a constant satisfying $J+D>0$. J indicates the strength of interband coupling constants. The degenerate eigenvalue is given by

$$\lambda = J + D. \tag{6.9}$$

θ and φ are Euler angles in the range of $0 \le \theta \le \pi$ and $0 \le \varphi \le 2\pi$.

The eigenvectors of the matrix in Eq. (6.8) that correspond to the eigenvalue $\lambda = J + D$ are

$$\begin{pmatrix} \cos\theta\cos\varphi \\ \cos\theta\sin\varphi \\ -\sin\theta \end{pmatrix}, \quad \begin{pmatrix} \sin\varphi \\ -\cos\varphi \\ 0 \end{pmatrix}. \tag{6.10}$$

When $\varphi = \pi/4$, they are

$$\begin{pmatrix} e^{i\phi} \\ e^{-i\phi} \\ 1 \end{pmatrix}, \quad \begin{pmatrix} e^{-i\phi} \\ e^{i\phi} \\ 1 \end{pmatrix}, \tag{6.11}$$

where the angle ϕ satisfies $\cos\phi = -1/(\sqrt{2}\tan\theta)$, and we assume that $|\sqrt{2}\tan\theta| \ge 1$. In this way the time-reversal symmetry broken state is obtained from the gap equation. If we set $\varphi = \pi/4$ and $\tan\theta = \sqrt{2}$, we obtain the case with three equivalent interband pairing interactions, that is, the three equivalent bands [17]. In this case we have $\phi = 2\pi/3$, and thus the phase difference between order parameters of three bands is $2\pi/3$.

7. Fractional Quantum Flux Vortices

7.1. Three-Band Superconductor

In this section let us discuss 1/3-quantized flux vorties in three-band superconductors. The Ginzburg–Landau free energy for a three-band superconductor is

$$\begin{aligned} f &= \sum_j \alpha_j \rho_j^2 + \frac{1}{2}\sum_j \beta_j \rho_j^4 - 2\gamma_{12}\rho_1\rho_2\cos(\theta_1-\theta_2) \\ &- 2\gamma_{23}\rho_2\rho_3\cos(\theta_2-\theta_3) - 2\gamma_{31}\rho_3\rho_1\cos(\theta_3-\theta_1) \end{aligned}$$

$$- \sum_j K_j \rho_j e^{-i\theta_j} \left(\nabla + i \frac{2\pi}{\phi_0} \mathbf{A} \right)^2 (\rho_j e^{i\theta_j}) + \frac{1}{8\pi} \mathbf{H}^2. \tag{7.1}$$

Here the order parameters are written as

$$\psi_j = \rho_j e^{i\theta_j}, \tag{7.2}$$

where $\rho_j = |\psi_j|$ is a real quantity. Since we focus on the role of phases of the order parameters, we assume that

$$\rho_j = \rho, \;\; K_j = K, \tag{7.3}$$

and define new phase variables

$$\Phi = \theta_1 + \theta_2 + \theta_3, \;\; \varphi_1 = \theta_1 - \theta_2, \;\; \varphi_2 = \theta_2 - \theta_3. \tag{7.4}$$

Then the free energy density is

$$\begin{aligned} f &= 3\alpha\rho^2 + \frac{3}{2}\beta\rho^4 - 2\rho^2 \left[\gamma_{12}\cos(\varphi_1) + \gamma_{23}\cos(\varphi_2) + \gamma_{31}\cos(\varphi_1+\varphi_2) \right] \\ &- 3K\rho\nabla^2\rho + 3K\frac{4\pi^2}{\phi_0^2}\rho^2\mathbf{A}^2 + \frac{1}{3}K\rho^2(\nabla\Phi)^2 + K\frac{4\pi}{\phi_0}\rho^2\mathbf{A}\cdot\nabla\Phi + \frac{1}{8\pi}\mathbf{H}^2 \\ &+ \frac{1}{3}K\rho^2\left[(\nabla\varphi_1)^2 + (\nabla\varphi_2)^2 + (\nabla(\varphi_1+\varphi_2))^2 \right], \end{aligned} \tag{7.5}$$

where $\alpha = (1/3)\sum_j \alpha_j$ and $\beta = (1/3)\sum_j \beta_j$. The stationary conditions with respect to φ_j lead to

$$\gamma_{12}\sin\varphi_1 + \gamma_{31}\sin(\varphi_1+\varphi_2) - \frac{1}{3}K\left[\nabla^2\varphi_1 + \nabla^2(\varphi_1+\varphi_2)\right] = 0, \tag{7.6}$$

$$\gamma_{23}\sin\varphi_1 + \gamma_{31}\sin(\varphi_1+\varphi_2) - \frac{1}{3}K\left[\nabla^2\varphi_2 + \nabla^2(\varphi_1+\varphi_2)\right] = 0. \tag{7.7}$$

7.2. Phase Potential and Chiral States

In this section we examine the ground state of the system with the potential

$$\begin{aligned} V = &- 2\gamma_{12}\rho_1\rho_2\cos(\varphi_1) - 2\gamma_{23}\rho_2\rho_3\cos(\varphi_2) \\ &- 2\gamma_{31}\rho_3\rho_1\cos(\varphi_1+\varphi_2). \end{aligned} \tag{7.8}$$

If we set $\varphi_3 = \theta_3 - \theta_1$, we have $\varphi_1 + \varphi_2 + \varphi_3 = 0 \pmod{2\pi}$. The minimum of this potential is dependent on the signs of the coefficients $\gamma_{ij}\rho_i\rho_j$ of the Josephson terms. We define $\Gamma_1 = -2\gamma_{12}\rho_1\rho_2$, $\Gamma_2 = -2\gamma_{23}\rho_2\rho_3$, and $\Gamma_3 = -2\gamma_{31}\rho_3\rho_1$. The potential is written as

$$V = \Gamma_1 \cos(\varphi_1) + \Gamma_2 \cos(\varphi_2) + \Gamma_3 \cos(\varphi_3). \tag{7.9}$$

We assume that the absolute values $|\Gamma_i|$ are almost equal in magnitude. Then there are four cases to be examined as shown in Table 1. When all the Γ_i are negative, we have the minimum at $\varphi_1 = \varphi_2 = \varphi_3 = 0$ (Case I). If we change the sign of Γ_3, this produces a frustration effect and φ_i take fractional values. For example, when all the $|\Gamma_i|$ are equal, we have a minimum at $(\varphi_1, \varphi_2, \varphi_3) = (\pi/3, \pi/3, 4\pi/3)$. In this state the order parameters are complex and thus the time reversal symmetry is broken. The case IV also exhibits a similar state with fractional values of φ_i. If all the $|\Gamma_i|$ are the same, the ground state is at $(\varphi_1, \varphi_2, \varphi_3) = (2\pi/3, 2\pi/3, 2\pi/3)$ (Figure 3).

7.3. Double Sine-Gordon Equation and Kinks

We consider the solution for $\gamma_{12} = \gamma_{23}$. In this case we have a solution with $\varphi_1 = \varphi_2 \equiv \varphi$. The variable φ satisfies the double sine-Gordon equation,

$$K\nabla^2\varphi - \gamma_{12}\sin\varphi - \gamma_{31}\sin(2\varphi) = 0. \tag{7.10}$$

Let us consider the kink solution of the one-dimensional double sine-Gordon equation. The energy functional is

$$E = \int \left[\frac{1}{2} K_0 \left(\frac{d\varphi}{dx} \right)^2 + V(\varphi) \right] dx, \tag{7.11}$$

where $K_0 = 2K\rho^2$ and the potential V is

$$V(\varphi) = V_0 \left(\cos\varphi + \frac{u}{2} \cos(2\varphi) \right). \tag{7.12}$$

We defined $V_0 = -\gamma_{12}\rho^2$ and $u = \gamma_{31}/\gamma_{12}$. There are two cases to be examined: (1) $\gamma_{12} < 0$ and (2) $\gamma_{12} > 0$. We show the classification of the double sine-Gordon model in Table 2.

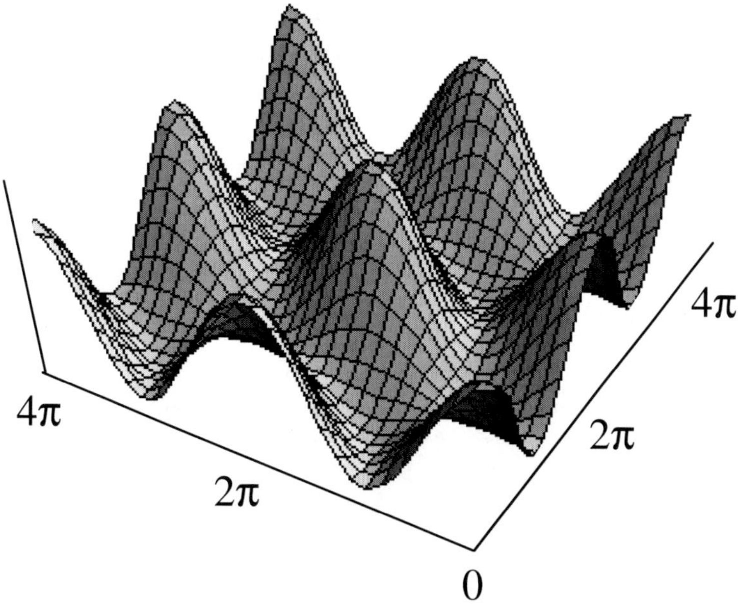

Figure 3. Contour map of V for $\Gamma_1 = \Gamma_2 = \Gamma_3 > 0$. Black and white dots indicate minima of the potential V. Dotted line is the path in the valley connecting two minima.

First consider the case (1) $V_0 > 0$. The potential $V(\varphi)$ has a minimum at $\varphi = \varphi_0 \equiv \cos^{-1}(-1/(2u))$ if $u > 1/2$:

$$V(\varphi_0) = V_0\left(-\frac{1}{4u} - \frac{u}{2}\right). \tag{7.13}$$

For $u \leq 1/2$, we have a minimum at $\varphi = \pi$:

$$V(\pi) = V_0\left(-1 + \frac{u}{2}\right). \tag{7.14}$$

In the case $u > 1/2$ we have a chiral state at $\varphi = \varphi_0$ and a kink solution that travels from one minimum to the other minimum. The double sine-Gordon equation

$$\frac{d^2\varphi}{dx^2} = -\frac{V_0}{K_0}(\sin\varphi + u\sin(2\varphi)), \tag{7.15}$$

Table 2. Classification of the ground state of the potential V of the Josephson interactions. φ_i in the Cases III and IV are for $|\Gamma_1| = |\Gamma_2| = |\Gamma_3|$

	Γ_1	Γ_2	Γ_3	φ_1	φ_2	φ_3	
Case I	$-$	$-$	$-$	0	0	0	
Case II	$+$	$+$	$-$	π	π	0	
Case III	$-$	$-$	$+$	$\pi/3$	$\pi/3$	$4\pi/3$	chiral state
Case IV	$+$	$+$	$+$	$2\pi/3$	$2\pi/3$	$2\pi/3$	chiral state

can be integrated for the boundary conditions $d\varphi/dx \to 0$ and $\varphi \to \varphi_+$ as $x \to \infty$ and $d\varphi/dx \to 0$ and $\varphi \to -\varphi_+$ as $x \to -\infty$. Here φ_+ is a solution of $\cos(\varphi) = -1/(2u)$ in the range $-\pi < \varphi < \pi$. Since the equation above is integrated as

$$\left(\frac{d\varphi}{dx}\right)^2 = \frac{2V_0}{K_0}\left(\cos\varphi + \frac{u}{2}\cos(2\varphi) + \frac{1+2u^2}{4u}\right), \tag{7.16}$$

we obtain the kink solution as The kink structure is shown in Figure 4.

In the case $V_0 > 0$ and $u \le 1/2$, there are minima at $\varphi = \pi \pmod{2\pi}$ in the potential V. Thus, we have a 2π-kink solution in this case. The boundary conditions should be $\varphi \to \pi$ as $x \to \infty$ and $\varphi \to -\pi$ as $x \to -\infty$, or vice versa. Since we obtain

$$\left(\frac{d\varphi}{dx}\right)^2 = 4\frac{V_0}{K_0}(1-2u)\cos^2\left(\frac{\varphi}{2}\right)\left(1 + \frac{2u}{1-2u}\cos^2\left(\frac{\varphi}{2}\right)\right), \tag{7.17}$$

the kink solution is given by

$$\varphi(x) = \cos^{-1}\left(1 - \frac{2\sinh^2(rx)}{\cosh^2(rx) - 2u}\right), \tag{7.18}$$

where $r = \sqrt{V_0(1-2u)/K_0}$.

Second, let us consider the case (2) $V_0 < 0$. There are minima at $\varphi = 0 \pmod{2\pi}$ for $u > -1/2$, and thus the 2π kink solution exists. We obtain

$$\varphi(x) = \cos^{-1}\left(\frac{2\sinh^2(sx)}{\cosh^2(sx) + 2u} - 1\right), \tag{7.19}$$

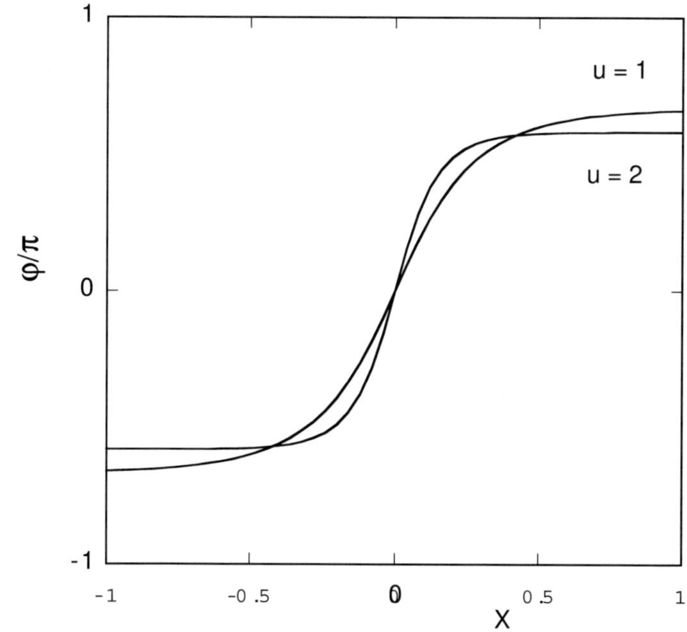

Figure 4. φ as a function of $X = \sqrt{2V_0/K_0}x$ for $V_0 > 0$ and $u = 1, 2$. Fractional kink structure is shown. For $u = 1$, $\varphi(-\infty) = -2\pi/3$ and $\varphi(\infty) = 2\pi/3$ which we call the $4\pi/3$ kink.

where $s = \sqrt{|V_0|(1+2u)/K_0}$. The kinks in this case are presented in Figure 5. For large u, the kink shows a characteristic at $x = 0$ because the potential has a local minimum at $\varphi = 0$ for $u > 1/2$. We have a possibility to find some specific features in the excited state due to this anomaly. For $u < -1/2$ we have a fractional-π kink that is given by

$$\varphi(x) = \tan^{-1}\left(\frac{1+2|u|t(x)}{\sqrt{4u^2-1}}\right) - \tan^{-1}\left(\frac{1+2|u|/t(x)}{\sqrt{4u^2-1}}\right), \tag{7.20}$$

where

$$t(x) = \exp\left(\sqrt{2\frac{|V_0|}{K_0}|u|\left(1-\frac{1}{4u^2}\right)}x\right). \tag{7.21}$$

For $u = -1$, this chiral solution satisfies the boundary condition that $\varphi \to -\pi/3$ as $x \to -\infty$ and $\varphi \to \pi/3$ as $x \to \infty$ as shown in Figure 6.

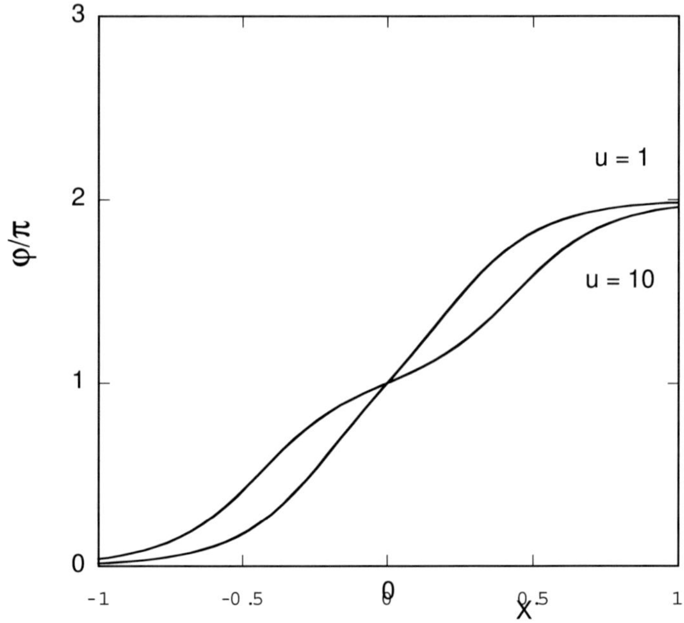

Figure 5. φ as a function of $X = \sqrt{(u+1/2)(2|V_0|/K_0)}x$ for $V_0 < 0$ and $u = 1$ and 10. This shows 2π-kink structure. For large u, the kink shows a saddle-like structure at $x = 0$. This is because the potential V has a local minimum at $\varphi = \pi$ for $u > 1/2$.

7.4. Energy of Kinks

We call the double sine-Gordon model the chiral sine-Gordon model when the conditions $V_0 > 0$ and $u > 1/2$ or $V_0 < 0$ and $u < -1/2$ hold. The energy functional for the one-dimensional system $-L/2 < x < L/2$ is

$$E = \int_{-L/2}^{L/2} \left[\frac{1}{2} K_0 \left(\frac{d\varphi}{dx} \right)^2 + V(\varphi) \right] dx. \tag{7.22}$$

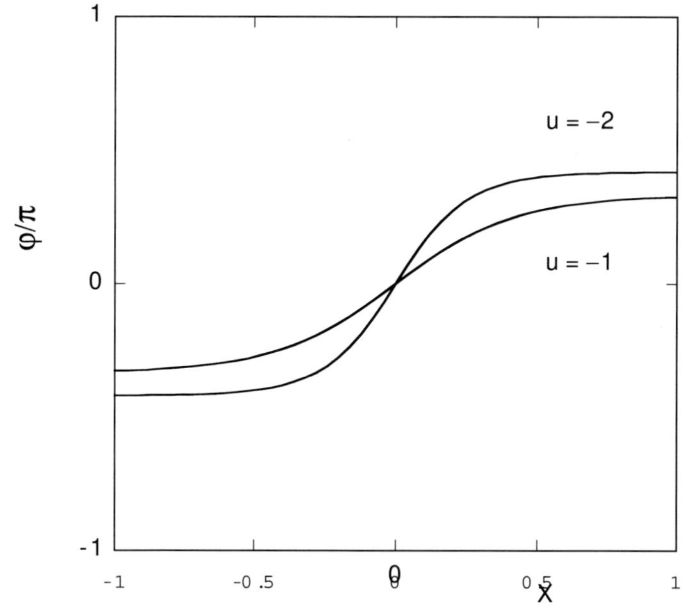

Figure 6. φ as a function of $X = \sqrt{2|V_0|/K_0}x$ for $V_0 < 0$ and $u = -1$ and -2. This shows 2π-kink structure. For $u = -1$, $\varphi(-\infty) = -\pi/3$ and $\varphi(\infty) = \pi/3$.

We assume that L is sufficiently large. If φ satisfies the stationary condition $\delta E/\delta\varphi = 0$, we obtain $K_0 d^2\varphi/dx^2 = dV(\varphi)/d\varphi$. From this we have

$$\frac{1}{2}K_0\left(\frac{d\varphi}{dx}\right)^2 = V(\varphi) - C_0, \tag{7.23}$$

where $C_0 = V(\varphi_0)$ with the condition that $d\varphi/dx \to 0$ as $\varphi \to \varphi_0$. Then the energy is

$$E = C_0 L + \sqrt{2K_0}\int_{\varphi_-}^{\varphi_+}\sqrt{V(\varphi) - C_0}\,d\varphi, \tag{7.24}$$

where $\varphi_+ = \varphi(L/2)$ and $\varphi_- = \varphi(-L/2)$, and both satisfy $d\varphi/dx \to 0$ as $\varphi \to \varphi_\pm$. For $V_0 > 0$ and $u > 1/2$, we obtain the energy of fractional-π kink state as

$$E_{f-kink} = 2\sqrt{2K_0V_0u}\left[\sqrt{1-\frac{1}{4u^2}} + \frac{1}{2u}\cos^{-1}\left(-\frac{1}{2u}\right)\right] + C_0 L, \tag{7.25}$$

Table 3. Classification of the double sine-Gordon model with the potential $V(\varphi) = V_0(\cos(\varphi) + (u/2)\cos(2\varphi))$. $V(\varphi)$ has minima at $\varphi = \varphi_0 \pmod{2\pi}$

V_0	u	φ_0	kink	
$V_0 > 0$	$u > 1/2$	$\cos^{-1}(-1/(2u))$	fractional-π kink	chiral
$V_0 > 0$	$u < 1/2$	π	2π-kink	
$V_0 < 0$	$u > -1/2$	0	2π-kink	
$V_0 < 0$	$u < -1/2$	$\cos^{-1}(-1/(2u))$	fractional-π kink	chiral

where $C_0 = V(\varphi_0) = V(\varphi_+) = V(\varphi_-) = -V_0(1/(4u) + u/2)$. This coincides with the energy obtained by substituting the kink solution directly to the energy functional. The energy of the 2π-kink for $V_0 < 0$ and $u > 0$ is

$$
\begin{aligned}
E_{2\pi-kink} &= C_1 L + 4\sqrt{K_0 V_0}\Bigg[\sqrt{1+2u} \\
&- \frac{1}{2\sqrt{2u}} \log\left|\left(1 - \frac{2u}{1+2u}\right) \Big/ \left(1 + \frac{2u}{1+2u}\right)\right|\Bigg], \qquad (7.26)
\end{aligned}
$$

where $C_1 = -|V_0|(1 + u/2)$.

Let us consider the case $V_0 > 0$ and $u > 1/2$. By using the inequality $a^2 + b^2 \geq 2|ab|$ for real a and b, we obtain

$$
\begin{aligned}
E &= \int \left[\frac{1}{2} K_0 \left(\frac{d\varphi}{dx}\right)^2 + V_0\left(\cos\varphi + \frac{u}{2}\cos(2\varphi)\right)\right] dx \\
&= C_0 L + \int \left[\frac{1}{2} K_0 \left(\frac{d\varphi}{dx}\right)^2 + V_0 u \left(\cos\varphi + \frac{1}{2u}\right)^2\right] dx \\
&\geq C_0 L + 2\sqrt{\frac{1}{2} K_0 V_0 u} \int \left|\frac{d\varphi}{dx}\left(\cos\varphi + \frac{1}{2u}\right)\right| dx \\
&= \sqrt{2K_0 V_0 u}\left[\sin\varphi_+ - \sin\varphi_- + \frac{1}{2u}(\varphi_+ - \varphi_-)\right] + C_0 L, \qquad (7.27)
\end{aligned}
$$

for the one-kink solution that satisfies $d\varphi/dx \geq 0$ and $\cos\varphi + 1/(2u) \geq 0$. In the case of one fractional-π kink shown above, we obtain

$$E \geq \sqrt{2K_0V_0u}\left[2\sqrt{1-\frac{1}{4u^2}}+\frac{1}{u}\cos^{-1}\left(-\frac{1}{2u}\right)\right]+C_0L. \tag{7.28}$$

where we adopt that $0 \leq \cos^{-1}\left(-\frac{1}{2u}\right) \leq \pi$. The lower bound of the energy coincides with the energy in Eq. (7.25). Here, we define the conserved current

$$J^{\mu} = \frac{1}{2A}\varepsilon^{\mu\nu}\partial_{\nu}\varphi, \tag{7.29}$$

with the charge

$$Q = \int_{-\infty}^{\infty} J^0(x)\,dx = \frac{1}{2A}\left[\varphi(x=\infty)-\varphi(x=-\infty)\right], \tag{7.30}$$

where $\varepsilon^{\mu\nu}$ is the antisymmetric symbol and $x^0 = t, x^1 = x$. A is the normalization constant defined by $A = \cos^{-1}(-1/(2u))$ with the value in the range $0 \leq A \leq \pi$. $\partial_{\mu}J^{\mu} = 0$ follows immediately from the antisymmetric symbol $\varepsilon^{\mu\nu}$ and does not depend on the equation of motion. The kink has $Q = 1$, and an antikink with $Q = -1$ exists that has a configuration with $\varphi(-\infty) = A$ and $\varphi(\infty) = -A$. If we assume that $\cos\varphi + 1/(2u) > 0$ for the kink solution, the energy inequality is written as

$$E \geq \operatorname{sign}Q\sqrt{2K_0V_0u}\,(\sin\varphi_+ - \sin\varphi_-) + \sqrt{\frac{2K_0V_0}{u}}A|Q| + C_0L. \tag{7.31}$$

This is an inequality of Bogomol'nyi type. For the kink satisfying $\cos\varphi + 1/(2u) < 0$, the inequality is

$$E \geq -\left[\operatorname{sign}Q\sqrt{2K_0V_0u}(\sin\varphi_+ - \sin\varphi_-) + \sqrt{\frac{2K_0V_0}{u}}A|Q|\right] + C_0L. \tag{7.32}$$

8. Fractional Vortices and Baryonic Bound States

In general, there are solutions of vortices with fractional quantum flux in multi-band superconductors. Kinks in the space of phase variables θ_j play a central role for the existence of fractional flux vortices as in the case of half-quantum flux vortices.

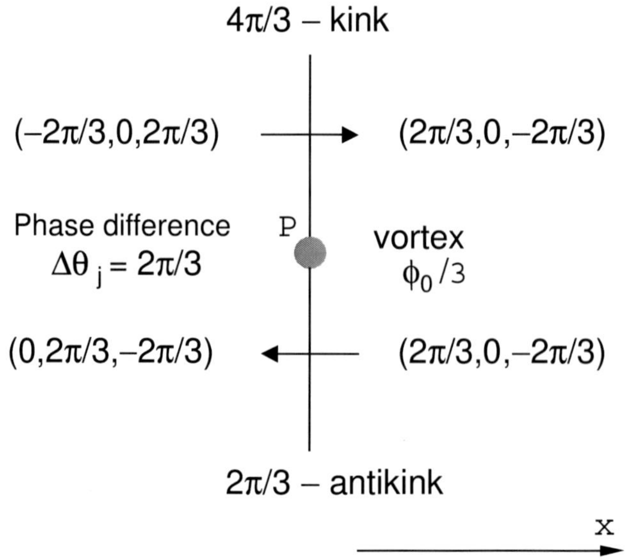

Figure 7. Kink, antikink and a fractional flux vortex for $\Gamma_1 = \Gamma_2 = \Gamma_3 > 0$. The vortex is at the point P with flux $\phi_0/3$ where ϕ_0 is the flux quantum. We start from $(\theta_1, \theta_2, \theta_3) = (-2\pi/3, 0, 2\pi/3)$ to reach $(0, 2\pi/3, -2\pi/3)$ (modulo 2π) through the $4\pi/3$-kink and $2\pi/3$-antikink. $\varphi_1 = \theta_1 - \theta_2$ goes from $-2\pi/3$ to $2\pi/3$ crossing the $4\pi/3$-kink, and φ_1 goes from $2\pi/3$ to $4\pi/3 \equiv -2\pi/3$ (mod 2π) through the $2\pi/3$-kink.

In three-band superconductors, the fractional-flux vortex exists in the chiral case as well as the non-chiral case. Since we have the fractional-π kink in the chiral state (Cases III and IV), the new types of vortices with fractional flux quanta exist on a domain wall of the kink [18–20]. The kink considered in the previous section is a one-dimensional structure in superconductors. There are many types of kinks connecting two minima of the potential in three-band superconductors.

Let us discuss the fractional vortices in the three-band model here. Suppose that two kinks, one is a kink and the other is an antikink, intersect at a point P as shown in Figure 7 in a two-dimensional xy-plane. If a vortex exists along the z axis just at the point P, the vortex should have a fractional flux quantum so that the phase change around the point P is 2π. For $V_0 > 0$ and $u > 1/2$, this

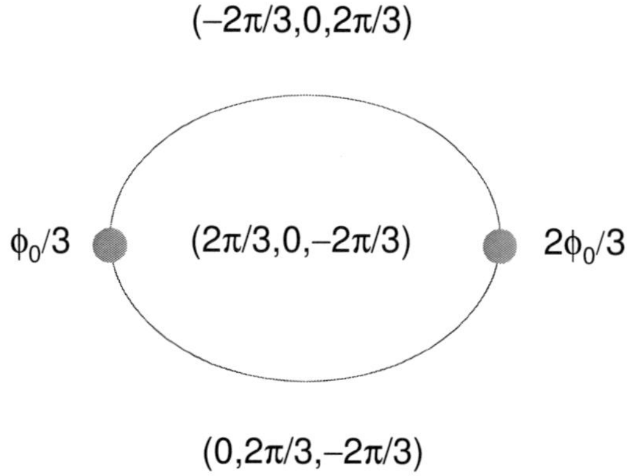

Figure 8. Two-vortex bound state with line singularities in the time-reversal symmetry broken state. The phase variables θ_i ($i = 1,2,3$) have singularities that are described by kinks in Figure 4. The total flux is ϕ_0. Topologically, the flux $2\phi_0/3$ is equivalent to $-\phi_0/3$. Thus, this state corresponds to the meson under the duality transformation between charge and magnetic flux.

is shown schematically in Figure 7. We set the phases of the order parameters $(\theta_1, \theta_2, \theta_3) = (-2\pi/3, 0, 2\pi/3)$ in some region. After crossing the $4\pi/3$ kink, they become $(2\pi/3, 0, -2\pi/3)$ where the phase variables φ_1 and φ_2 change from $-2\pi/3$ to $2\pi/3$. If there is also a domain wall of an antikink that starts from the point P as in Figure 7, we have the phases $(0, 2\pi/3, -2\pi/3)$ after we cross the antikink. Here, we obtain the phase difference between the initial and final states (see Figure 7). In this case, the vortex that is located through the point P along the z axis should have a fractional flux quantum $\phi_0/3$. Thus, in the chiral region of three-band superconductors, the existence of fractional vortices is easily concluded in this way.

In the three-band model, the fractional vortex has two line singularities (kinks) in the phases of the gap function as shown in Figure 7. From Figure 7, we have a two-vortex bound state as presented in Figure 8 in the chiral state. Two vortices form a 'molecule' by two kinks. This state may have lower energy than the vortex state with quantum flux ϕ_0 since the magnetic energy $(5/9)\phi_0^2$ is smaller than ϕ_0^2 of the unit flux. The energy of kinks is proportional to the distance R between two fractional vortices if R is large. Thus, the attractive

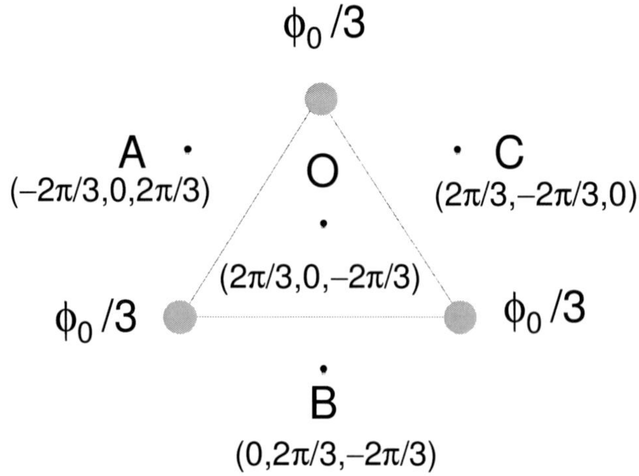

Figure 9. Three-vortex bound state with line singularities in the time-reversal symmetry broken state. Each vortex has $\phi_0/3$ and the total flux is ϕ_0. The phase variables θ_i $(i = 1,2,3)$ have singularities that are fractional-π kinks. One can read one $\phi_0/3$ as $-2\phi_0/3$. Hence, this state corresponds to the neutron.

interaction works between them if R is sufficiently large.

Three-vortex bound states are also formulated: they are shown in Figures 9, 10 and 11. The first two figures indicate bound states in the time-reversal symmetry broken state. The last one is for the unbroken state [21]. These states correspond to baryons if we regard the magnetic flux as charge.

9. Confinement and Deconfinement

In quantum Chromodynamics (QCD), quarks do not appear as asymptotic states. Quarks are constitutes of nucleons and mesons and cannot be separated arbitrary far from the rest of the constituents. This is the hypothesis of confinement [22]. The interquark potential is expected to be a linearly increasing function so that quarks are confined. To understand such a potential, one may consider a thin flux tube between separated quarks. The energy of the flux tube is proportional to the length of the tube. This is an analogy to magnetic vortices in type-II superconductors where magnetic fields penetrate into superconductors

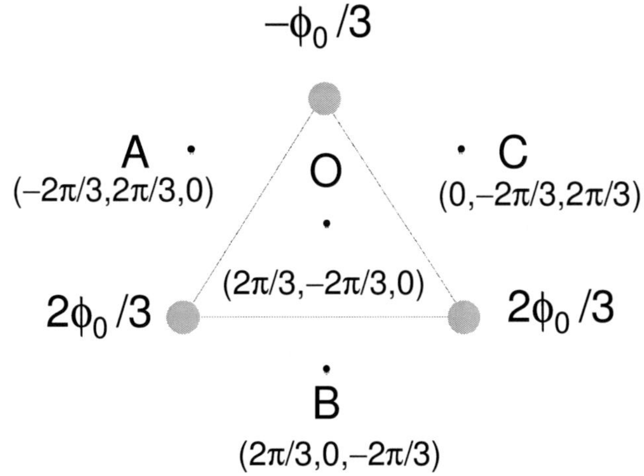

Figure 10. Three-vortex bound state with line singularities in the time-reversal symmetry broken state. This state corresponds to the proton.

with quantized flux. Because the energy of vortex per unit length is constant, the total energy of vortex is proportional to its length. The confinement of quarks can be understood as confinement in dual superconductor model as a result of the dual to Meissner effect.

Let us consider a pair of half-quantum flux vortices discussed in Section 5. Because two half-flux vortices are connected by a line of singularity (namely, domain wall), the potential energy between them is proportional to the separation of two vortices,

$$V(r) = \sigma r, \tag{9.1}$$

for large r where σ is a constant proportional to $\sqrt{K_0V_0}$. This has an analogy to quarks in a charged pion because of a linear confinement potential. The linear potential in a superconductor originates from the kink in the phase space of the order parameters. The kink of half-quantum flux vortex is a defect in two-dimensional space and this attracts two separated vortices. When two half-quantum vortices become separated far away each other, it is more energetically favorable for a new half-quantum pair to appear, than to let the kink to extend further.

A vortex with unit flux ϕ_0 has the energy proportional to ϕ_0^2, and a half-

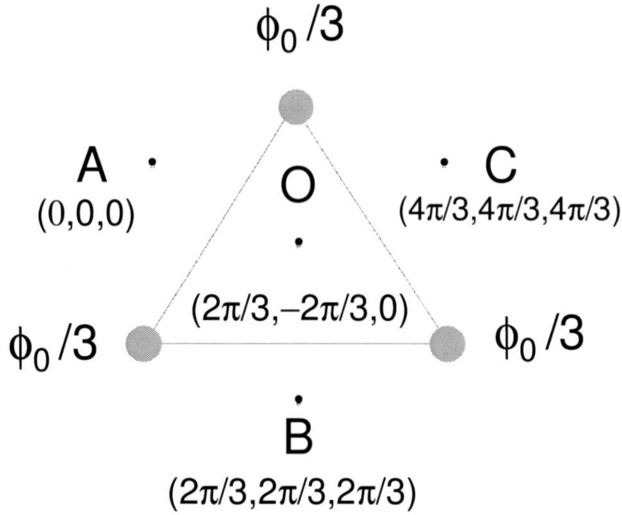

Figure 11. Three-vortex bound state with line singularities in the time-reversal symmetric state. Each vortex has $\phi_0/3$ and the total flux is ϕ_0. In this state, the region including the point O has higher energy.

quantum flux has the magnetic energy which is proportional to $\phi_0^2/4$. Hence, the separation L between two half-flux vortices is determined by the energy balance,

$$\frac{1}{2}\phi_0^2 = AL, \tag{9.2}$$

for some constant A. Because $A \propto \sqrt{V_0}$ and V_0 is proportional to the strength of Josephson coupling γ, L increases as the Josephson coupling γ decreases.

Goryo et al proposed a mechanism of deconfinement of two vortices due to an entropy effect [23]. They calculated the entropy and estimated the temperature above which two fractional vortices are deconfined. The obtained deconfinement temperature T_{dec} is very close to T_c, and thus free fractional vortices hardly exist at $T < T_c$.

Summary

There are a number of interesting analogies between ideas in particle physics and corresponding one in condensed matter physics. Heisenberg considered a microscopic theory of ferromagnets using quantum mechanics, and later employed an analogy with ferromagnets to construct a unified theory of elementary particles [24,25]. Nambu imported the idea of spontaneous symmetry breaking in condensed-matter physics to particle theory to account for the mass of fermions, especially nucleons. He first notices the resemblance between the Dirac equation in field theory and the gap equation in superconductivity theory.

There are now other analogies between particle physics and physics of superconductivity. A half-quantum flux vortex can be regarded as a monopole, and fractionally quantized-flux vortices bear a striking resemblance to quarks. Two half-quantum vortices form a pair with a linearly increasing potential, which corresponds to mesons. They are not allowed to become separated far away each other for $T < T_c$. In a three-band superconductor, three fractional-flux vortices can form a bound state with the total flux ϕ_0 or 0, which corresponds to the proton or neutron, respectively, after the dual transformation between charge and magnetic flux.

References

[1] Ginzburg, V. L.; Landau, L. D. *J. Exptl. Theor. Phys. (JETP)* 1950, 20, 1064.

[2] Nambu, Y. Proceedings of the Midwest Conference on Theoretical Physics, Purdue University, p. 1 (1960); *Broken Symmetry Selected Papers of Y. Nambu*; Eguchi, T.; Nishijima, K.; Eds.; World Scientific: Singapore, 1995.

[3] Nambu, Y.; Jona-Lasinio, G. *Phys. Rev.* 1961, 122, 345.

[4] Nambu, Y.; Jona-Lasinio, G. *Phys. Rev.* 1961, 124, 246.

[5] Bardeen, J.; Cooper, L.; Schrieffer, R. *Phys. Rev.* 1957, 106, 162.

[6] Bogoliubov, N. N. *J. Exptl. Theor. Phys.* 1958, 34, 58, 73; translated in *Soviet Phys. JETP* 1958, 7, 41, 51.

[7] Nambu, Y. *Phys. Rev.* 1960, 117, 648.

[8] Gor'kov, L. P. *J. Exptl. Theor. Phys.* 1959, 36, 1918; translated in *Soviet Phys. JETP* 1959, 9, 1364.

[9] Suhl, H.; Matthias, B. T.; Walker, L. R. *Phys. Rev. Lett.* 1959, 3, 552.

[10] Geilikman, B. T.; Zaitsev, R. O.; Kresin, V. Z. *Soviet Phys.– Solid State* 1967, 9, 642.

[11] 't Hooft, G.; *Nucl. Phys.* 1981, 190, 455.

[12] Izyumov, Yu. A.; Laptev, V. M. *Phase Transitions* 1990, 20, 95.

[13] Tanaka, Y. *Phys. Rev. Lett.* 2002, 88, 017002.

[14] Babaev, E. *Phys. Rev. Lett.* 2002, 89, 067001.

[15] Leggett, A. J. *Prog. Theor. Phys.* 1966, 36, 901.

[16] Kobayashi, M.; Maskawa, T. *Prog. Theor. Phys.* 1973, 49, 652.

[17] Stanev, V.; Tesanovic, Z. *Phys. Rev. B* 2010, 81, 134522.

[18] Tanaka, Y.; Yanagisawa, T. *J. Phys. Soc. Jpn.* 2010, 79, 114706.

[19] Tanaka, Y.; Yanagisawa, T. *Solid State Commun.* 2010, 150, 1980.

[20] Yanagisawa, T.; Tanaka, Y.; Hase, I.; Yamaji, K. *J. Phys. Soc. Jpn.* 2012, 81, 024712 (http://arxiv.org/abs/1202.5400).

[21] Nitta, M.; Eto, M.; Fujimori, T.; Ohashi, K. 2010, arXiv:1011.2552.

[22] Wilson, K. G. *Phys. Rev. D* 1974, 14, 2455.

[23] Goryo, J.; Soma, S.; Matsukawa, H. *Europhys. Lett.* 2007, 80, 17002.

[24] Nambu, Y. In *Physics in the 21st Century*; Kikkawa, K.; Kunitomo, H.; Otsubo, H.; Eds.; World Scientific: Singapore, 1997.

[25] Heisenberg, W.; *Introduction to the Unified Theory of Elementary Particles*; John Wiley and Sons Ltd.: London and New York, 1966.

INDEX

B

C

D

E

F

G

H

I

J

K

L

M

N

O

P

Q

R

S

T

U

V

W

Y